Forensic Biomechanics and Human Injury

Criminal and Civil Applications –
An Engineering Approach

Forensic Biomechanics and Human Injury

Criminal and Civil Applications – An Engineering Approach

Harold Franck
Advanced Engineering Associates
St. Petersburg, Florida, USA

Darren Franck
Advanced Engineering Associates
Charleston, West Virginia, USA

CRC Press
Taylor & Francis Group
Boca Raton London New York

CRC Press is an imprint of the
Taylor & Francis Group, an **informa** business

CRC Press
Taylor & Francis Group
6000 Broken Sound Parkway NW, Suite 300
Boca Raton, FL 33487-2742

First issued in paperback 2020

© 2016 by Taylor & Francis Group, LLC
CRC Press is an imprint of Taylor & Francis Group, an Informa business

No claim to original U.S. Government works

ISBN-13: 978-1-4822-5883-7 (hbk)
ISBN-13: 978-0-367-77886-6 (pbk)

Visit the Taylor & Francis Web site at
http://www.taylorandfrancis.com

and the CRC Press Web site at
http://www.crcpress.com

Contents

Symbols and Units

In this section, we list the symbols we use throughout the book for quick reference. We have also listed common units used to perform calculations. Note that in many instances units are in the English system (pound, foot, second) and other units are in the metric system (kilogram, meter, second) also known as SI. The section on conversion factors allows for the easy conversion between the two systems. Unfortunately, in the United States, an attempt to convert all units to the metric system has not succeeded to follow the rest of the world. Consequently, some conversions are necessary but are relatively simple.

Greeks Letters

φ	Golden ratio of the Greeks = 1.618034...
x_n	Fibonacci numbers = 0, 1, 1, 2, 3, 5, 8, 13,...
K_e	Kinetic energy (ft lb)
w	Weight (lb or kg)
g	Acceleration due to gravity (ft/s^2 or m/s^2)
v	Velocity (ft/s or m/s)
V	Volume (ft^3 or m^3 or gal or L)
a	Acceleration (ft/s^2 or m/s^2)
d	Distance (ft or m)
F	Force (lb or kg)
m	Mass (kg or slugs or lb)
x, y, z	Distance in the particular direction (ft or m)
σ	Stress
A	Area (ft^2 or m^2)
ε	Strain
δ	Deformation
ρ	Density
l or L	Length (ft or m)
E	Young's modulus
τ	Shearing stress
t	Thickness
P_n	Normal load
P_t	Transverse load
v	Poisson's ratio

e	Dilation
p	Pressure
k	Bulk modulus
T	Torque
ρ	Radial distance
J	Polar moment of inertia
G	Shear modulus
φ	Angle of twist
I	Area moment of inertia
S	Section modulus
M	Moment
Q	First moment of inertia
V	Shear in the section
r	Radial arm
μ	Coefficient of friction
K	Effective length factor
S_x^p	Sensitivity of P to x
V_{xj}	Variability of x
x_{am}	Arithmetic mean
x_{gm}	Geometric mean
$\boldsymbol{a_t}$	Unit vector in the tangential direction
$\boldsymbol{a_r}$	Unit vector in the radial direction

Note: Vectors may be represented by a **boldface** symbol.

Preface

There are many reasons why people write books. Our intent for writing this book is threefold. The first reason is to bring together information about the strength of biological materials from a variety of sources in a concise, comprehensive manner that is readily accessible to the reader or practitioner. Most of the information on the characteristics of biological materials, from an engineering standpoint, is difficult to find unless the reader has a fairly comprehensive library or does a significant amount of research. A considerable amount of the information exists in an outdated published form. Some of the information must be gleaned from animal studies simply because it is not available and tests have not been performed on humans. The reason for the lack of data on some human properties is simply that those studies could not be performed on humans. There is insufficient concise information on the subject in one text that is readily accessible to the investigator without performing extensive searches. In order to find some of the information, multiple sources from biology, anatomy, strength of materials, and medicine must be researched.

Let us turn for a moment to the topic of animal studies and why they are appropriate in our context. Animal studies have been utilized to determine the characteristics of scientific scrutiny in many forms, including the response to chemicals and carcinogens to which it would be inappropriate to subject humans. In our context, we simply include animal studies to correlate the similarities of the physical properties of various tissues to the human anatomy. When data for humans are not available, the correlations we have included afford the investigator a reasonable range of values of the characteristics of the particular tissue.

The second reason for writing this book is to educate the public, and in particular, engineers, attorneys, and judges, on the methodologies utilized to compute the forces, stresses, and energies required to produce injury to the human body. The calculations are actually quite simple. The trick is to determine how to apply the equations to the human body with relatively accurate dimensions and material strength characteristics to properly model the forces and stresses on the particular body part resulting from a particular incident. The intent is also to argue the simple fact that biomechanical engineers are the most capable individuals to perform these calculations. Again, let us emphasize that biomechanics is not a science of diagnosis but rather a science of causality and correlation to injury.

The third reason is to allow a relatively inexperienced person a method to perform the calculations, which may be especially true for a criminal investigator who is given the task to assign blame for the incident. The criminal investigator may need to perform some basic calculations to assess whether an incident should be included in the category of criminality or not. As an example, it may be alleged that the injuries suffered in a homicide case are due to a fall down a staircase. Some basic calculations may yield a completely different scenario in which it may be shown that the injuries could not have been created by a fall but rather by multiple blunt trauma to the body. Another example of a criminal case may involve a broken tibia alleged to be caused by the fall of a female. Calculations may show the injury to have actually resulted from physical abuse by a boyfriend or husband. A third example may involve shaken baby syndrome by a guardian in which it was alleged that the baby simply fell on the floor. Of course, in the case of injury or death, medical imaging or an autopsy would validate such findings of the forensic biomechanical investigator. All the parts of the puzzle must fit together in a cohesive manner. In this context, we have produced a variety of examples to compute the damaging events. Furthermore, we have included some calculations into which the investigator can easily plug typical values of known data and calculate the unknown. The pertinent equations are already included in the calculations with some typical values. Here, the investigator only needs to change certain parameters to fit his application. The parameters that may need to be entered to fit the particular case are found in the applicable sections of the book.

By writing this book in a relatively simple manner, without the complex medical terminology and anatomical description found in most treatises, we hope to educate and elucidate practitioners on the subject of injury to humans. Some mathematics and terminology are required to properly comprehend the subject matter, which is especially true of medical terms that confound most of us. We hope that the medical terminology encountered from the description by doctors will mostly be found in this text without the need to perform further searches. We have, however, simplified the material substantially and allowed the reader the opportunity to investigate how and under what conditions humans get hurt. Much of the medical terminology on the minor components of the human anatomy has been disregarded because we are looking at the major or most common types of trauma that occur. For example, we do not address the forces that may cause a broken toe because that type of injury can occur simply by stubbing the toe against an object while walking.

Acknowledgments

We express our gratitude to our families, including Maria, Amanda, and Janice for all the help and support received during the process of producing this book. Without their help, this book could not have been written. We also express our gratitude to the staff of CRC Press, in particular to Mark Listewnik, who encouraged the development of the topics in this treatise.

Authors

Harold Franck founded Advanced Engineering Associates Inc. (AEA) in 1989 and has a home office in Charleston, West Virginia and a satellite office in St. Petersburg, Florida. Since that time, he has been involved in thousands of forensic engineering investigations involving vehicle accident reconstruction, origin and cause fire investigations, and electrical incidents. He is a registered professional engineer in the states of West Virginia, Michigan, Ohio, Virginia, Kentucky, and Florida. Harold received his BSEE from the West Virginia Institute of Technology (WVIT) in 1970 and an MSEE from West Virginia University in 1973. He began his academic work as an instructor of electrical technology at WVIT in 1970, eventually becoming associate professor of electrical engineering in 1981 before retiring in 1990. Harold has presented and attended various courses and seminars through the years, lists many publications, and has completed two books:

> *Forensic Engineering Fundamentals*, 2012. Coauthor: Darren Franck. Boca Raton, Florida. CRC Press, Taylor & Francis Group.
> *Mathematical Methods for Accident Reconstruction*, 2010. Coauthor: Darren Franck. Boca Raton, Florida. CRC Press, Taylor & Francis Group.

Darren Franck is a registered professional engineer in the state of West Virginia. He is president of AEA, Inc. and has been with the company since 1995. Darren received his BSCE from Virginia Polytechnic Institute and State University in 1999 and an MSME from the Georgia Institute of Technology in 2005. Darren's areas of expertise include forensic engineering investigations, structural analysis and design, accident reconstruction, computer-aided design, and 3D animations. He has been involved in various consulting, construction management, and design activities throughout West Virginia. Darren is also coauthor of two the books listed earlier.

Introduction

1

Humans are prone to injury by a variety of methods with no limitations. The three most common methods include disease, war, and accidents. The first method of injury that is produced by disease is essentially the purview of the medical expert. The proclivity of the human condition to the various maladies associated with disease is not the subject of this book. Injuries produced by war and conflict are well documented and can be categorized as being caused by extremely violent acts. These acts usually involve the detonation of explosives, weapons of mass destruction, or sophisticated armaments that are designed to cause permanent injury and more likely death. Again, we are not concerned with these types of injuries although their results may be similar to the types of injury that we encounter in this book.

A third method by which humans are injured is through what is commonly referred to as an accident. Accidents come in many forms and include slips and falls, car accidents, or may be equipment or humanly produced. For clarity, we choose to refer to this type of injury as an incident rather than an accident. The main reason for the reference of an incident is because these injuries may not be accidental in nature. They may be caused by a variety of events some of which cannot be categorized as accidental. Additionally, they may not be injurious but are claimed to be so. Others may actually be predicted by the events that precede them. Actions or events may be put in place that determine a root cause for the event or incident. A simple example of such an event is a football player who suffers a head injury, returns to the game, and is then diagnosed with a closed head injury. Some of the most common incidents involve humans in perfectly acceptable activities such as sports. Some of these injuries may be caused by sudden, forceful actions in the activity being performed or they may be produced by continued cyclic activity that degrades the affected tissue. Knee, shoulder, and elbow injuries associated with sports such as football, baseball, and tennis come to mind. These injuries may be sudden as when a soccer player changes direction rapidly or they may be produced by cyclic loading resulting from swinging a tennis racket over many years. Irrespective of the type of injury, forensic engineers performing biomechanical calculations are mainly concerned with the forces and accelerations that the activity or incident produced. They are interested in the strength properties of the affected material. These forces can then be correlated to the strength of the affected tissue structures in order to determine if the energies were sufficient to cause the injury.

It is well understood that the potential for injury increases with increasing age. The human system, as with any other system, is affected by wear and tear. Barring the effects of disease on the human body, the human structures that make us up will eventually wear out. Age-specific maladies include the degradation of the joints, muscles, bones, tendons, and ligaments. The strength of these biological materials decreases with age and should always be considered when analysis is performed for the potential of injury by a particular incident. Whether a person is young or old, when the human structures affected are subject to excessive loading beyond their limits, the structures are subject to failure. This failure may take many forms including stress, strain, rupture, or catastrophic failure as in the form of a compound fracture. Compound fractures are those that cause the bones to protrude through the skin and are deemed the most severe and have a greater potential to cause death via infection.

From a historical perspective, it should be pointed out that biomechanical calculations were not carried out before the twentieth century although Newton's laws were well-known for at least two centuries prior. The reason for this late development in calculating and quantifying the forces and energies required to injure biological tissue is threefold. First, investigators were not aware that biological tissue was subject to the same laws that applied to nonanimate structures such as wood, steel, cement, and others. These biological structures were simply not studied. The science of biomechanics and strength of materials had not emerged and matured. Second, the properties of biological materials were not known. Stresses, strains, and loading of biological structures were not known. Humans and animals had not been subjected to the forces and accelerations in order to quantify the effects of severe loading. Before the advent of mechanized motion, humans and animals were not stressed by significant velocities. These velocities only emerged as a result of the machines that were primarily invented in the twentieth century. Prior to the industrial revolution, injuries to humans were produced by jumps, falls, weapons, war, and human mayhem. Technology had not advanced to the point where velocities and accelerations produced by machines affected a large portion of the population. Additionally, the surfaces that impacted humans when exposed to falls and slips were generally much softer. For example, there is a marked difference when falling from a horse at 20 mph onto a grassy field when compared to striking a wall riding a motorcycle at the same speed. Although concussions have been medically known for a long time, the propensity of the sports, actions, and machines available today have a much greater effect on this type of injury. Our mechanized world has produced the third element in the proliferation and scientific study of biomechanical loading.

This book has several chapters that encompass the biomechanical forensic engineering field. Before we continue with a discussion of the chapters in the book, it is important to differentiate the term biomechanical from biomedical. The term biomedical is the purview of professionals trained in medicine. Medically trained professionals, whether they are doctors, nurses, or physical therapists, have a completely different perspective on injury to humans. They are trained to diagnose and correlate injury and disease. They are trained to repair and mollify injury. They are not necessarily trained to determine the correlation between injury and the forces required to produce that injury although their experience allows them to correlate certain activities to the degradation of biological structures. Their experience and training also allows them to determine the detrimental effects of these structures due to aging or repeated loading. For example, older individuals are prone to degradation of their joints, such as their knees, whereas young people are generally not subject to these conditions. Similarly, baseball pitchers are commonly subject to shoulder and elbow degradation as a result of their particular sport. In this context, medically trained personnel are the most qualified individuals to determine the correlation between activity, age, and injury.

In contrast, biomechanical engineers are generally not trained to correlate the previously mentioned injury and causation or diagnosis. Their training is geared toward the forces, energies, velocities, and accelerations that will cause biological materials to fail. The tools of analysis for a biomechanical engineer involve the application of Newton's laws and Strength of Materials principles to the structures that make up the human body. In terms of the physical structures, bones, muscles, tendons, ligaments, and organs, and other animal structures, the terms of forces, accelerations, stresses, and strains all behave in predictable and known ways. That is, these biological materials have properties that can be studied within the context of their strength, stress, and strain. The literature is replete with the physical properties of these materials. Consequently, the scientific literature on the onset for injury to biological materials is well understood, studied, and documented. Biomechanical engineers can perform calculations using the known scientific literature about the strength of tissues and correlate the potential for injury to a particular event or incident. They are generally not trained to diagnose injury or disease unless they are trained in the medical sciences. Biomechanical engineers are normally not doctors or nurses; however, they are certainly capable of understanding human anatomy and terminology. Consequently, they are well suited to perform the calculations that correlate injury to the human body to the forces that produce injury.

In many jurisdictions, the sitting judge will not allow a forensic biomechanical engineer to testify to the potential or correlation to injury as a result of a particular event. These judges only allow medically trained professionals to testify to the correlation between injury or the potential for injury.

This condition promulgated by the sitting judge may actually be inaccurate or erroneous because the medical expert may or may not be trained in the mechanics of injury. In many instances, the medical expert simply correlates the potential for injury to the statements of the involved parties which may or may not be accurate. Simply put, the involved parties are not disinterested participants and may have a stake or concern with the outcome of the diagnosis. Someone with a preexisting condition, such as a herniated disc, may be involved in a minor collision and seek treatment based on the preexisting condition, claiming that the injury was produced by the minor accident. In fact, this accident may not be the cause of the injury which can be determined by the forensic biomechanical engineer through the calculation of the forces, accelerations, and the strength of the affected biological materials.

In actuality, biomechanical engineers are eminently qualified to make such calculations but may not be permitted by the court to render such opinions. The necessary and sufficient conditions required to make the calculations and render the opinion are quite basic—these are a basic knowledge of the mechanical laws of physics, an understanding of strength of materials science, basic dimensions of the affected tissues, and their physical properties in term of stresses, strains, and ultimate strength. The fact that the judge may not allow the opinion of the biomechanical engineer is beyond the scope of the purview of the biomechanical engineer and simply depends on the enlightenment of the presiding judge. Why some judges choose to allow medically trained personnel to opine on the cause of an injury who do not possess the background in biomechanics is not understood by the authors but is recognized. It is simply a fact of life and part of an imperfect judicial system.

In reality, the most qualified individuals to perform biomechanical calculations are professionals involved in biomechanics. These individuals are engineers, scientists, physicists, and other similarly trained professionals who are involved in testing and calculating the necessary forces that injure human tissue. The ideal biomechanical professional would have training in medicine and engineering. Although these individuals do exist, the nature of each of these fields makes them a rare commodity. The intensity of training involved in both fields makes it very difficult to achieve this level of competence and knowledge. In the opinion of the authors, training in engineering with knowledge of the strength of biological materials suffices to render opinions concerning injury to humans.

Historical Developments

Before the twentieth century biomechanical calculations were not performed. There was simply no interest or need for the quantification of the energies involved in the injury to humans. This lack of interest changed with the events leading to the Second World War. In particular, Germany and Japan undertook an effort to study the effects of forces that injured humans. Some of these efforts were in fact cruel, immoral, and without regard to the sanctity of human life. Some of the experiments were barbarous and have been well documented historically. However, cruel this type of experimentation was, it did yield some valuable information on the strength of biological materials.

A second source of human tolerance to injury came from the U.S. space program and the X-plane tests performed at various facilities by the Air Force. Live subjects were placed in a variety of conditions that tested the limits of tolerance of the human body. These studies revealed just how strong and tolerant humans are to forces and accelerations. These studies were necessary to determine if humans could be placed in rockets and accelerated to the escape velocity necessary to reach orbit and then return to earth. Typical orbital velocities attained by a variety of propulsion methods to reach space produce accelerations of approximately 8–9 times the acceleration on gravity, commonly referred to as 8–9 g's. Actually, humans have been subjected to accelerations in excess of 30 g's under controlled conditions without injury to the human test subjects. Keep in mind that these individuals were physically fit, and strict control was used in the experimentation. It should not be assumed that all humans are capable of withstanding 30 g's or more. However, it is very reasonable that normal humans can tolerate accelerations of 8–9 g's. In fact, tests with human subjects of all shapes, sizes, and physical conditions place the onset of injury at about 12 g's. The scientific literature indicates that soft tissue injuries begin at about 12 g's and that by 17 g's the possibility of soft tissue injury is most certain.

The third source of information on human tolerance comes from academia and professional societies. These sources have been most active in the past 50 years since the mass proliferation of the automobile and the mayhem that it causes. A significant amount of testing on humans and cadavers has taken place and is still very active. Testing on cadavers is performed with the consent of the deceased and family members and is conducted in a proper and dignified manner. In fact, restraint systems were developed as a consequence of this type of testing. Seat belts and air bags were the result of this testing, and the design of these systems is based on the scientifically known tolerance to injury. As an example, we might consider the deployment of the air bag. The deployment is activated by a speed change of approximately 12–14 mph depending on the vehicle and the manufacturer. This speed change correlates to an acceleration of between 10 and 12 g's. Why was this value chosen? Surely not by a lucky guess but rather as a result of numerous tests that yielded the tolerance of humans to noninjurious forces and accelerations. If humans were injured at accelerations of approximately 4 g's and the air bag systems did not deploy until a level of 12 g's was reached, then the air bag systems would be deemed ineffectual. Alternatively, if air bag systems are deployed at 4 g's produced by a typical medium bump on the road or a broach on a curb, then drivers would be very unhappy with their vehicles. It is conceivable that numerous law suits would be filed for the unnecessary deployment of the air bag system. It is necessary to note that there are experts who claim that humans can essentially be injured at any velocity. If that were the case, it can be reasonably argued that we, as an evolutionary animal, would not have made much progress in the course of our development as a species. How could humans have evolved if minor forces would cause injury and subsequent death? Scientific testing and calculations simply do not support these claims. That is not to say that there is not great variability in the strength of human tissue. Of course there is, but there are upper and lower bounds as with any material. When biomechanical calculations are carried out, this variability in the strength of human tissue is always taken into account. Not to do so is misleading and, to a certain degree, incompetent because the physical properties of that individual are never known to an exact degree. To err on the side of caution, the calculations are always carried out over the range of known scientific possibilities for the strength of these biological materials. This point will be explained further in other sections of the book as we carry out examples of the possibility for injury.

This book contains information on the strength of biological materials of other animals. The question that may arise is why such information is provided. From an evolutionary standpoint, it is well-known and understood that we, as a species, are not very different from our nonhuman relatives. The development of the human species can be traced to our earliest roots for approximately 500 million years to the Cambrian explosion. The Cambrian explosion signifies the rapid development, diversity, and similarity in the life forms that evolved during this period of earth's history. In a simple narrative, it can be said that rudimentary life forms underwent significant changes in their procession toward more complex life forms and developed very similar characteristics that have continued to this day. These similar characteristics include the development of four limbs, a backbone, a tail, a circulatory system, and a gastrointestinal digestive system to name a few. Although in outward appearances, most life forms have very distinct characteristics, the basic parts are generally similar under strict scrutiny.

Testing of biological materials of humans is somewhat limited by ethical concerns. We simply cannot take a living human and break a leg, puncture a lung, or rupture a tendon. Such activity is considered inhuman and barbarous although such activity has been known to have occurred. Our society still has problems dealing with assisted suicide in the case of impending death in order to alleviate suffering. We, as a species, attempt to cling to life by any and all medical procedures that maintain some semblance to life.

The authors take no sides in this very complex debate but recognize that the criteria applied to other animals are different. Most of us consider euthanasia the humane approach when other animals, such as our pets, face suffering or impending death. Consequently, some of the testing on other animal subjects does not adhere to the same strict requirements imposed on humans.

Additionally, not all human testing of biological materials has been carried out. Some testing of biological materials of other animal species has been performed so that these tests can be correlated to human tissue. A statistical analysis of human versus other animal tissue reveals the similarity in the strength of the materials. When the strength of human biological materials is not known or determined, similar animal materials may be used.

This correlation between human and animal materials can invoke controversy as outlined by the Daubert versus Merrell Dow Pharmaceuticals case approximately 20 years ago. For those readers not familiar with this case, we offer a very condensed version of the case.

In 1993, the U.S. Supreme Court determined the standard for the admission of expert testimony in the federal court system. The *Daubert Court* articulated what is known as the *Daubert Standard,* which overturned the Federal Rules of Evidence as outlined in the *Frye Standard*. The *Frye Standard* stemmed from a 1923 case, which only allowed evidence to be admitted in court if the evidence was in consort with the general acceptance in the field. In 1975, Congress adopted the Federal Rules of Evidence, which included the Frye Standard as part of federal common law. Rule 702 of the Federal Rules of Evidence did not make expert testimony admissible depending on general acceptance. In the *Daubert* case, the plaintiffs contended that the drug Bendectin caused birth defects in humans based on animal studies, which had not gained acceptance in the scientific community.

As a result of these cases, the new standards that govern expert testimony must meet three provisions. First, the testimony must be based on scientific knowledge. Second, the testimony must assist the jury or the judge in understanding the evidence within a pertinent context. Third, the testimony must be scientifically valid, tested, subjected to scientific peer review, published, generally accepted, and rate of error known. A more concise set of guidelines for the new standard as promulgated by Rule 702 of the Federal Rules of Evidence can be summarized where the reliability of the evidence must meet a nonexclusive four part test. This test is as follows:

1. Can the theory or technique be tested?
2. Has it been subjected to peer review and publication?
3. Is there a known or potential for error?
4. Is there general acceptance in the scientific community similar to the Frye Standard?

The validity of the use of animal biological materials and their properties as a substitute for human biological materials is fully developed and correlated in Chapter 8, which is accomplished by correlating the materials and applying statistical deviations between the two to prove their relevance.

Beyond this introductory chapter, the book has an additional 14 chapters, applicable federal standards, and a bibliography. The chapters are grouped into four categories. The first category includes the need for biomechanical analysis and an explanation of how injuries occur relative to the activities of the human subjects. This category includes Chapters 3 through 5. The second category includes elements of anatomy, terminology, and physical characteristics of biological materials. These topics are covered in Chapters 6 through 10. The third category includes the applicable static and dynamic equations used in the analysis. Chapters 9, 11, and 14 deal with these subjects. The fourth category includes protective structures and standards that are covered in Chapters 14 and 15.

Chapter 2 deals with the court system and testimony of the expert. Preparation for depositions and trial along with presentations by the expert are covered. The demeanor and believability when testifying are the most crucial elements when the expert is on the witness stand. This chapter may be read at any time. However, the other chapters should be read in sequence for clarity of presentation. Chapter 15 on the Federal Standards is included because, in many cases, proving that a standard has been violated is prima fascia evidence. References to applicable sections are made throughout the book.

On a stylistic note, the authors are aware that not all experts, scientists, and engineers are male. However, the clumsy constructions "his/her," "she/he," and their variations have been avoided for clarity's sake. It should be assumed that a masculine pronoun is not meant to refer only to the male gender. Mankind includes all of us.

Court System and Testimony

2

The practicing biomechanical forensic engineer has to be intimately aware and knowledgeable of the court system in the United States. Although there is some variability in the court system from state to state, most of the elements of the court system are the same. There is also some variability in the testimony and evidence that an expert is allowed to hear and know in a particular case. There may also be variability in the system from county to county in a particular state. Additionally, within a county, the litigating parties may agree to certain elements that are not necessarily within the norm of that particular jurisdiction. An attempt to describe all the different court systems is not the purpose or intent in this chapter. Some of the most common variations are discussed.

Our basic system of justice is predicated upon the English system of justice that dates back to the Magna Carta in 1215. The Magna Carta was influenced by King Henry I who specified that a limitation would be placed on certain of his powers. This limit resulted from and was influenced by the Charter of Liberties of 1100. After 1215 the Magna Carta underwent many modifications, and by the time that the early settlers reached New England, it was influential in their interpretation of law. As a direct consequence, the Magna Carta played an influential role in the development of the Constitution of the United States. The Englishmen who colonized America were greatly influenced by the Magna Carta when they established their charters. These included the Virginia Charter, the Maryland Charter, and the Massachusetts Bay Company Charter. These early colonists' interpretation of the Magna Carta was anachronistic in that they believed that it guaranteed trial by jury and habeas corpus and was a fundamental law. Habeas corpus essentially means "you have the body" and relates to the direction of a prison warden to produce the person so that the court can determine the status and legality of the custody of the prisoner. The court order concerning the production of the habeas corpus is known as a writ.

Consequently, the framers of the U.S. Constitution designed the legal system in the manner of English common law and the philosophy of John Locke who is regarded as the *Father of Classical Liberalism*. The significant influence of John Locke on the development of the founding documents of the United States cannot be underestimated. Locke was a true believer in empirical data in consort with the underpinnings of the beliefs of Francis Bacon. John Locke was extremely influential in social contract theory, liberal theory, republicanism, and their significance to the Declaration of Independence, the Bill of Rights, and the U.S. Constitution. Locke's ideas on liberty, social contract, property, price theory, monetary thought, political theory, religious beliefs, and the concept of self influenced many of the founding fathers of this country.

There are three main levels of the court system in the United States. At the top level there is the Supreme Court of the United States. Under this level of the court system, there are sublevels of the federal court system referred to as federal district courts. At the next level, there are the individual state supreme courts. These state courts are also made up of lower levels and include the district and county courts of the states. Litigation generally begins at the lowest level, which is the county courts and then the litigation proceeds from the state district courts to the state supreme courts. Thereafter, the litigation may work its way up through the federal court system and eventually end up at the Supreme Court of the United States.

The Supreme Court is made up of eight justices appointed by the president and confirmed by the senate. The chief justice is the ninth member, who is appointed and confirmed in the same manner as the other justices. The other justices are referred to as associate justices. Their appointment is for life and consequently may sway the opinions of the court in significant directions toward liberalism or conservatism. The cases heard by the Supreme Court are limited each year by the justices and proceed from federal or state courts. In other words, the appeals may arise from federal district courts or from state supreme courts. There are 13 federal district circuit courts of appeal. These federal appeal courts arise from 94 federal judicial districts of which 12 are regional circuit courts. The thirteenth district has jurisdiction over specialized cases relative to international trade and the Court of Federal Claims. There are 50 state supreme courts. Figure 2.1 shows the U.S. district courts. The numbers corresponding to the colored sections of the map in Figure 2.1 represent the various regional circuit courts of the United States.

Individual cases that may involve a practicing forensic biomechanical engineer would most probably not involve the opinions of the expert unless the case involves constitutional issues as with the Daubert case. The Supreme Court is the final arbiter of federal constitutional law and was established by the Constitution in 1789 in Article Three. These cases involve attorneys who present oral arguments on the merits of the case, and the forensic expert would not testify but his opinions may be cited.

The 50 state supreme courts are, in their function, essentially identical to the federal supreme court system in that they normally hear appeals and do not make any finding of facts. When the state supreme courts find errors in a particular state decision, they will order the court to retry the case. The case is remanded to the lower court. The state supreme courts vary in their makeup depending on the particular constitution of the state in which they reside.

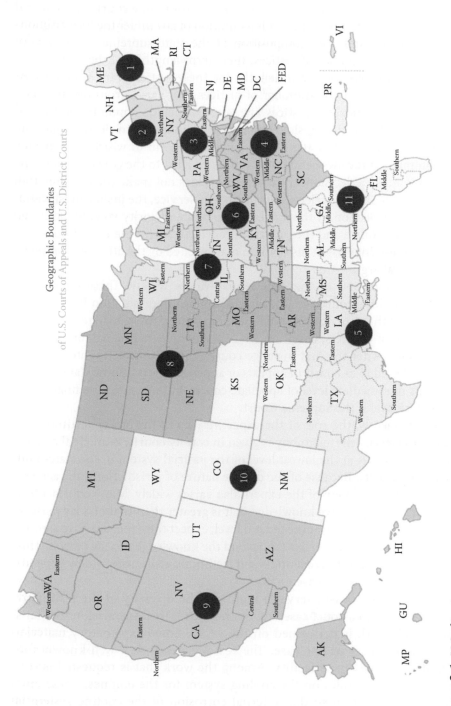

Figure 2.1 U.S. district courts.

These courts are distinct from federal courts in the same or similar geographic boundary. In some instances, appeals from these courts may end up in the U.S. Supreme Court if there is a question of law under the Constitution of the United States. The composition of the state supreme courts varies widely from the number of justices, their term length, and the method by which they are selected. Term lengths for these justices vary widely from 6 years to life terms. The number of justices also varies from five to nine. The modes by which the justices are selected include appointment by the governor, elected by the legislature, partisan or nonpartisan election, and the Missouri Plan. The Missouri Plan originated in 1940 in which a nonpartisan commission reviews candidates and sends a list to the governor who has 60 days to make a decision. If the governor does not make the decision, the commission makes the selection. After 1 year service, the justice must stand in a retention election. If the judge fails to win a majority vote, the judge is removed from office and the process is repeated. Again the expert does not testify in these courts, but his opinion may be cited by the attorneys presenting their legal arguments.

The composition of the state courts varies widely. These courts are divided into three courts, beginning at the general jurisdiction level, sometimes called the court of first instance. According to each state, they may be made up of districts, counties, circuits, superior, municipal, or supreme designations. After the first instance courts, there are usually intermediate courts, referred to as courts of appeals or appellate courts. Some states do not have this intermediate level of appellate courts, but most do. The final court at the state level is the Supreme Court.

Consequently, the role of the expert resides in his or her testimony at the district level. This testimony is given in county courts, municipal courts, and circuit courts at the lowest level of the judicial system in the states and the federal courts. Because of the diverse nature of jurisdictions, the acceptability of the testimony of the expert also varies widely. Many courts allow an expert to simply have knowledge that is greater than that of a lay person. Other courts, especially at the federal level, expect the expert to have significant knowledge and expertise far beyond the knowledge of a lay person. This scrutiny of the expert's credentials and testimony also varies widely depending on the nature of the case.

Some examples will serve to highlight the diversity of the testimony that is allowed in a court case. Let us choose a boat transmission that fails after repair work is performed on the vessel. Some explanatory material is necessary to set up the case. The vessel is taken to a well-known and regarded marine repair facility. Among the work that is requested is servicing the zinc anodes on the cooling system for the engines. These zinc anodes are sacrificial so that internal corrosion of the cooling system is minimized. In order to access the zincs on the transmission cooler, some disassembly of transmission components is necessary. In the process, one of the high-pressure transmission fluid hoses is damaged. When the vessel is recommissioned, the transmission fluid leaks out and damages the transmission. Cause and effect are clear in this case. The hydraulic fluid only leaked out after the vessel was brought in for servicing and the hose was damaged. Or was it? A claim could be made that excessive pressures developed in the transmission as a result of worn parts that burst the transmission hose. For the first scenario, the expertise required is little more than that of a competent mechanic knowledgeable in marine transmissions. For the second scenario, a marine transmission engineer may need to be called to bolster the arguments made. In this second scenario, the opinions and qualifications of the engineer may come into play if a Daubert challenge is imposed by one of the parties. The challenge may come in two forms. One form may be that the marine transmission engineer may never have studied transmission failures and is not cognizant of the various modes of failure. The other form may come from the application of failures to transmissions on large vessels and not smaller recreational vessels; in effect arguing that his opinions are based on studies performed on very different types of transmissions. Thus, his opinions are based on junk science. Additionally, the expert may not have been able to reference transmission failures of the particular kind in this vessel.

A more pertinent example for this book may come from expert testimony in an injury case. Let us set up this case as follows. On a dark night in a remote, unlit portion of the city, one vehicle allegedly rear-ends another at an intersection. The collision is relatively minor producing little damage to either vehicle. In fact, the collision may have actually not happened as described by the parties involved. Moreover, the reconstruction of the alleged collision raised physical data and produced calculations that were inconsistent with the collision. The reconstruction indicated as a worst case scenario that the speed changes experienced by the occupants of the vehicles were less than 5 mph. Based on the reconstruction, a biomechanical forensic engineer calculated the forces and accelerations experienced by the occupants from the collision. Based on these calculations and the known scientific data on injuries to humans, the expert concluded that the injuries claimed were inconsistent with the alleged collision. The injuries that were being claimed were limited to spinal bulging discs. The opposing expert, also trained in biomechanics, alleged that humans can be injured at any speed but cannot site any scientific studies to bolster his opinion because such evidence does not exist. The injured party was treated by a chiropractor whose opinion was based on the patients' word of the injury and its onset relative to the collision. The chiropractor neither performed any diagnostic testing nor referred the patient for such tests. The opposing side, cognizant of the unlikely nature of the injuries being claimed, referred the patient for x-rays and magnetic resonance imaging (MRI) testing. These tests revealed scoliosis of the spine with some disc bulging on the 65-year-old patient. The medically trained expert physician opined that the condition of the patient was due to age, a congenital condition, and was preexisting. Therefore, the question arises: Who should be allowed to testify and can challenges be made on the admissibility of testimony?

Let us first consider the opinions of the chiropractor. In some states, chiropractors are considered medical professionals and are called, in fact, doctors. Their opinions carry the same weight as the opinion of a medical doctor trained in orthopedics. The difference between the two specialties is in the depth and extent of the training that each of these professions require. In jurisdictions where both types of medical experts are allowed to offer opinions as to causation, the jury must decide which expert is more credible. It seems reasonable that a well-informed jury would select the opinions of the highly trained orthopedic medical doctor over that of the chiropractor. However, as a result of the complexities of litigation, the orthopedic doctor may not be able to testify. In that case, the jury may only hear the opinions of the chiropractor.

Let us turn to the biomechanical experts. First, assume that the biomechanical engineer was disqualified by the court to offer an opinion on the causation of the injuries claimed because he did not have a medical background. In that case, the causation would be predicated on the opinions of the medical specialist. If the orthopedic expert did not testify, then the jury would only hear the opinion of the chiropractor who based his analysis solely on the statements of the injured party. Such scenarios arise often in litigation, and juries hear alleged scientific evidence that is not based on facts and does not conform to the scientific principle. In the opinions of the authors, this conforms to junk science but is allowed in many courts in this country.

Second, assume that the court is progressive and allows biomechanical experts to testify as to cause and effect of injury to humans based on science. Both forensic biomechanical experts testify. One is stating that the injuries claimed are not consistent with the scientific literature, and the other is stating that injuries occur as a result of any velocity change or acceleration. Which expert will the jury believe? We hope the expert whose opinions are based on science and not speculation. These different scenarios are not imaginary, and all have occurred to the authors.

Role of the Expert

In this portion of the chapter, we hope we can explain the necessary elements that make up a successful expert witness. An expert witness must possess a variety of talents or assets. That is not to say that the perfect expert witness exists, but a combination of these elements to a reasonable degree are required. These elements, in our opinion, are as follows:

Qualifications
Technical expertise
Report writing ability
Conciseness of opinions
Scientific validity
Presentation and demeanor

Qualifications

The term biomechanical engineer explains very succinctly the qualifications of the expert. Traditionally, engineers are divided into four major categories: civil, chemical, electrical, and mechanical. Over the past 40 years, these four major divisions in engineering have been expanded within each of the divisions. For example, many civil engineers perform work in the environmental area so that civil engineering programs at universities have developed specialties in that area at the undergraduate level. The rapid development of computers have proliferated computer engineering programs emanating from electrical engineering departments. Other engineering programs at universities have developed a variety of options in engineering at the baccalaureate level and also at the masters and PhD levels. Although a few schools offer degrees in biomechanical engineering, most schools offer the option as a special application of the field of engineering and may fall under a department such as mechanical or electrical engineering. For most applications in the forensic area of biomechanical engineering, the level of knowledge of statics, dynamics, and strength of materials at the undergraduate engineering level suffices along with some rudimentary knowledge of anatomy and nomenclature. This knowledge can easily be obtained by taking a short course from an organization such as the Society of Automotive Engineers through continuing education.

Engineering means that science, physics, and mathematics are applied to describe events or phenomena. Engineering is an applied science. Biomechanics incorporates two disciplines, mechanics and biological material. Mechanics is fundamental to all fields of engineering. Biological materials are made up of living structures of a varied nature. These biological materials are subject to the same laws of physics as other materials such as steel, concrete, wood, and plastics, etc. Consequently, their physical properties, such as strength, strain limits, and general physical dimensions, are well known and understood. The limits of these materials with respect to injury and failure are also well known and can be calculated by the proper application of simple physical laws. Therefore, the qualifications of a biomechanical engineer can be classified as a person with engineering training and knowl-edge of anatomy and the properties of the biological tissues.

Technical Expertise

In order to perform biomechanical calculations, the expert needs to know the following principles: force, displacement, velocity, acceleration, area, volume, energy, stress, and strain. In engineering, knowing the principles involves applying the proper mathematical description of the situation and performing the calculations. If the expert does not perform calculations and show the underlying theoretical development, then the expert does not meet the basic criteria of engineering. Simply put, the technical expertise involves making the calculations to support the premise and opinion. In the practice of engineering according to the cannons of ethics, proposing an opinion without supplying the corresponding calculations is simply a dereliction of duty and subject to censure of some form by state engineer-ing boards.

Report Writing

Report writing is important because the audience of the report is very diverse, from insurance professionals to medical professionals to attorneys and judges and to highly trained experts in biomechanics. This diversity of readers of the expert's report requires many challenges for the author of the report. Let us look at each of the possible readers of the report that a forensic biomechanical engineer may write.

Consider first, the source of the assignment. It may be from an attor-ney who represents an injured client or it may come from a defense attorney or insurance professional who has an interest in the case. For example, the case may be as a result of an alleged failure of the safety system of a vehicle such as an airbag or a seat belt. The case may involve a walkway safety issue or an industrial accident where a worker was injured. The forensic biome-chanical engineer may represent either side of the case. Irrespective of the side being represented, the expert shall not be an advocate but a searcher for facts. Opinions must be based on science, and these opinions in the form of a report must be understood by the nontechnically trained reader.

The opinions must be simply and clearly written so that the lay reader understands the events of the case and where responsibility for the event lies. The reader should not misunderstand this statement. By stating "where the responsibility for the event lies," the authors in no way are implying that the biomechanical engineer decides fault. The engineer makes calculations that relate the injury to the event and describe the potential for that injury based on the known science. Forensic biomechanical experts, as with most experts, tend to write reports designed to be read by other experts. Consequently, the reports are filled with technical jargon, mathematical formulas, and complex theoretical arguments. Although there is a place for such verbosity, this type of analysis should be reserved for an appendix. In fact, several items of a report should come in the form of appendices. A report template that works well is outlined by the following:

I Title Page
This portion includes to whom the report is directed, the nature of the case with identifying numbers, and the entity that authored the case.

II Letter of Transmittal
A short letter is directed to the person who provided the assignment stating the nature of the assignment, the persons involved in the case, and signed by the originator of the report. In the case of an engineering analysis, the letter should be stamped to conform to professional engineering standards.

III Review of Findings
This is the main body of the report and includes what information was reviewed, what activities were undertaken, and what type and method of analysis was used to arrive at the conclusions.

IV Conclusions
This section summarizes the results of the analysis and cites applicable standards used in the report. The first four sections of the report are directed to the lay reader and not necessarily the opposing expert.

V Appendices
There may be any number of appendices that may include diagrams, graphs, photographs, analysis, references, and other supporting documentation. Most of this material would be directed toward other experts and not the lay reader although photographs and to a certain extent diagrams should be viewed by the lay reader.

VI Invoice
The invoice is an integral part of the report because it should reflect the date of the activity, the time spent on the activity, the rate of remuneration for each of the activities, any extra expenses (i.e., travel costs, photographs, etc.), and a grand total for the report.

This report outline is used by the authors and reveals their preference based on many years of experience. It should not be construed as the only form of a report, but many of its elements tend to answer questions that arise from the client and the opposing attorney at deposition or trial. The report not only allows the opposing side to have the answers to their questions but also to help the expert refresh his memory of the activities he undertook to formulate his opinions.

Many experts become verbose in their reports. We think this is a mistake because reports should be brief, to the point, concise, and well organized. Extraneous words simply confuse the issue and require explanations at deposition or trial. The expert does not want to answer "what did you mean by that statement" from opposing counsel. It is better to follow the writing style of Ernest Hemingway rather than that of Henry James.

On a final note with respect to report writing, many experts include their resume as part of the report. The reason for this practice is unclear, but may simply be included to increase the volume of the report. Content is what matters and not the number of pages in the report. Additionally, this practice is unnecessary because the client normally has a copy of the resume and the opposing side, if not familiar with the expert, will be asking for a copy in interrogatories.

Conciseness of Opinions

Many experts tend to hedge their bets by including all possible scenarios and instances related to the event. Opinions should only be based on the available physical evidence and not on scenarios unrelated to the case. Speculation should never enter into expert opinions. If a premise cannot be expressed with a reasonable amount of certainty, then it should not be considered. This leads to the question: What is reasonable? In engineering, the term within "engineering accuracy" is often used. This degree of accuracy in a calculation indicates that the value obtained should be between 5% and 10% of the actual value. Most of the time engineering calculations do not have an exact value because all of the parameters are not exactly known. For example, the ultimate tensile strength of upper thoracic human wet vertebrae for 20–39-year-olds is 0.37 ± 0.01 kg/mm². For 40–79-year-olds, it is 0.33 ± 0.02 kg/mm². The adult average is 0.34 kg/mm². What value should one use when making the calculation, let us say, for a 40-year-old individual? From the data listed, the highest value is 0.38 and the lowest value is 0.31. It could be argued that a 40-year-old could fit both age categories. So the average is 0.345 ± 0.035. The percent error is given as

$$\frac{0.345 \pm 0.035}{0.345} \times 100\% = 0.345 \pm 10\% \qquad (2.1)$$

We see from this example that actual values used in analysis perfectly conform to the reference of engineering accuracy. Furthermore, the computations performed in this manner allow for the variability and error rate to be expressed. Error rate should always be expressed in engineering forensic calculations. The error rate is easily expressed in tabular or graphical form.

Scientific Validity

Without question, values in a computation cannot be assumed without proper science-based valuation. Let us take the value for the example earlier. An expert may wish to argue that the 40-year-old is a woman who had a multitude of ailments so that the ultimate tensile strength of her vertebrae was much lower, say 0.20 kg/mm². Of course, the weaker the value, the less force and energy required to produce a failure. In a damaging event, a lower value would correspond to lower velocities and accelerations that the expert hopes will support his calculations and opinions. However, such actions are not conforming to any scientific validity. Such a calculation is paramount to using a value of 20,000 pounds per square inch (20 ksi) for the strength of steel instead of the usual value of 36,000 pounds per square inch (36 ksi). Such a calculation is simply incorrect, misleading, and without any scientific basis.

The expert may try to dance around the issue by stating that a lower value was used because of the feebleness of the individual. However, this is an incorrect application because the feebleness should not be explained by a decrease in the tensile strength, but rather by the degradation of the physical dimensions of the material. As with the steel example, for a section of steel that has corroded, the strength of the steel is not decreased in the computation, but rather the physical dimensions are adjusted by appropriately decreasing their size. The same procedure must be used when attempting to perform the biomechanical calculations on human tissue. Otherwise, the computation is incorrect. As with any proper engineering calculation, biomechanical calculations should take into account the variability in the physical dimensions and the variability in the properties of the materials. Only then can a proper assessment of the potential for injury be performed in consort with proper scientific and engineering validity. If a structural engineer attempted to perform a calculation using 20 ksi steel on a computation, he would produce incorrect results.

Presentation and Demeanor

There are two phases to this section. One phase of the presentation and demeanor is the report of the expert, which we hope has been covered in adequate detail. A well-written, concise, scientifically valid report that covers all the bases goes a long way to establish the credibility of the expert. The second phase involves the actual physical presentation of the person. We will discuss this phase in detail in this section.

The most important portion of this phase is the competence and knowledge of the expert. The expert must be very well prepared and in command of the science and mathematics. From a technical perspective, an expert who falters when asked to derive an equation or fumbles around technical issues simply appears as a poor witness. In depositions and even at trial, the expert is often asked to derive or write a mathematical expression used for his opinions. The expert should perform this task with ease and command of the underlying physics relative to basic principles. It is acceptable for the expert to refer to notes or references to give the values of obtuse constants but not of common constants such as the gravitational constant. However, when it comes to derivations of equations, the expert must have full command of the process from first principles to the end resulting equation used in the calculation. This command of the scientific equations is achieved through practice and bodes well for the credibility of the expert in the eyes of the opposing attorney, the trial judge, and, most importantly, the jury.

Finally, the expert must always stay calm even under duress when the opposing attorney insults or demeans the expert. One should never raise his voice or look away from the jury. Questions may be answered firmly, with conviction, but never in a confrontational tone. Always make the opposing attorney look like the bully and concede the points that are true. Never argue but concisely and truthfully answer all questions. Do not fall into the typical trap where the attorney says YES or NO is the only answer to a question. You are allowed to answer and explain your answer fully. Poise and self-confidence without aloofness serve the expert very well. It cannot be stressed enough that you should continually look at the jurors, not one juror, but all the jurors when you answer questions on the stand at trial. During a deposition, look directly at the opposing attorney when answering his questions. The attorney is sizing you up to note your competence, behavior, and credibility as a witness.

For most of us, public speaking or answering questions in a deposition or at trial is not natural, especially, when the tone is confrontational. Seldom is the tone from the opposition not confrontational. That is the nature of the court system. Remember that the opposing attorney is not concerned with the facts of the case but rather with representing his client. This approach may include introducing novel, unscientific theories and evidence not in accord with the facts. Unlike engineering and science, legal theories of a case are not based on provable, demonstrative, and repeatable facts. The best ally of the expert is technical competence and practice. Developing your demeanor when being asked confrontational questions is a matter of practice. Stay even toned and assured of your responses. Although we may not possess the natural tendencies to come across as a creditable witness, these skills can be acquired and should be practiced.

Depositions are almost always more combative on the part of the opposing attorney mainly because in a deposition there is no judge to caution the attorney. Consequently, some attorneys become very confrontational and accusatory especially in the tone that they affect. Since the deposition is

usually transcribed and not recorded, the attorney assumes that his intimi-
dating tone will not be reflected in the record. One way to soften this abuse
is to simply say words to the effect, "sir/madam, there is no reason to get mad
at me and yell. I will answer all your questions in a respectful manner just
as long as you are civil." Remember, in a deposition, the opposing attorney
wants to see how you handle pressure and will most likely be confrontational
to a certain extent. In a trial, the presiding judge will normally not allow the
opposing attorney to intimidate you. The best advice is to always stay cool
and do not lose your temper. Juries, especially, do not like attorneys who
harangue an expert witness. As an expert, you are not on trial, but are there
simply to relay the science and the facts. If you have the facts and the science
correct, there is not much that the opposing attorney can do.

At trial, the plaintiff normally presents his case followed by the defense.
You may be working either side of a case. Irrespective of the side you are
working for, you may be called to rebut the testimony of the opposing witness. Make it a point to not argue with the attorney and simply present the
facts. Concede the points where there is commonality between your opinion
and that of the other expert. Digress where your opinions are different and
backed by the science.

How Injuries Occur

3

Humans are very innovative when it comes to being injured. Most injuries are caused by human activity in one form or another, and very few can be classified as purely accidental. Consequently, as we have outlined previously, we refer to such events as incidents. The seven major categories of the types of injury to humans may be classified as either accidental, self-inflicted, inflicted by others, unforeseen events that cause injury, faulty equipment, defectively designed equipment, or events that can be predicted. Let us look at each of these classifications in greater detail in order to assign proper blame for the event or incident.

Accidents

As we have stressed in this book, we simply opine that accidents are rare events. According to Webster's dictionary, an accident is a sudden event that is not planned or intended that causes damage or injury and occurs by chance. It is further defined as an unforeseen and unplanned event or circumstance. According to this definition, an accident would be a very rare event indeed. With respect to injuries to humans, it is easily noted that so-called accidents would not have a clearly defined root cause because they are unforeseen. In the case of staged accidents, these events are indeed well planned with the intent to defraud the insurance carrier. Consequently, they should not be called accidents but rather one of the other categories that we have listed earlier. We would classify staged accidents as planned events that can be predicted and generally do not cause actual injury. The injuries claimed cannot be corroborated through proper medical evaluations and testing. If the event can be assigned a root cause, prediction, or chain of consequences, then it cannot be classified as an accident. Most incidents actually have a root cause that can be analyzed and often predicted in hindsight. One of the main reasons for analyzing events that injure people is to minimize their impact or to totally eliminate them, which is true for any of the types we have listed. People generally believe that accidents are the least likely events to occur as a type of injury. With those thoughts in mind let us delve a little further into the various types of events that we have outlined.

Self-Inflicted

Injuries that are classified as self-inflicted may take a variety of forms. For example, if a person drives too fast, loses control of the vehicle, wrecks, and is injured, then the injuries are self-inflicted. If a person uses a power saw in an inappropriate manner, in conflict with the instructions provided, and cuts a finger, then the injury is clearly self-inflicted. If a skier with little experience in complex terrain decides to take on a challenging downhill course and falls breaking a leg, then we opine that the injury is self-inflicted. There are a variety of injuries that may be classified as self-inflicted as a result of carelessness, incompetence, misjudgment, improper use of equipment, or lack of training to perform a certain task. On a more sinister note, some self-inflicted wounds or injuries may have an ulterior reason for their occurrence. This form of a self-inflicted injury may be perpetrated by a person who is committing a crime and for some reason wants to appear as a victim. The crime may involve civil deception or may be purely criminal in nature. Consequently, these persons may cause bodily harm to themselves in order to perpetrate some form of a deception. These types of self-inflicted injuries may involve striking objects, knives, or gunshots. Typically, these injuries are superficial, but the perpetrator claims loss of consciousness or incapacitation. Some basic calculations on the type and extent of the injury may reveal the deception. These calculations may involve a determination of the bullet path, the puncture of the skin with a sharp object, or a self-inflicted contusion to a part of the body. In many instances, the alleged injuries may not reveal themselves to physical examination when there is no apparent contusion, cut, abrasion, or physical semblance of injury. These types of injuries may not reveal themselves in x-rays or CT scans. Simply, the claimed injury may result from a desire by the perpetrator to avoid work, make fraudulent claims, or cover up a crime.

Inflicted by Others

The most obvious injury inflicted by other individuals is when one person attacks the other causing injury or death. These assaults may be inspired by self-defense or by malice. In that case, criminal actions may be taken by law enforcement and the legal system. In other cases, the injuries may be produced as a result of ignorance, accident, or carelessness. Sometimes, injury to other humans is caused by the inaction of certain parties. One example of this type of injury may involve the inaction of certain parties to guard individuals from being injured by not following recognized standards. Simple examples include the lack of machine guarding of equipment, uneven walkway surfaces, lack of electrical injury protection for workers, and the lack of fall protection of workers. For these types of injuries, there are a multitude of federal and state standards. These include the Code of Federal Regulations, International Building Codes, Engineering Standards, and a variety of Testing and Procedural Standards promulgated by various organizations. These are generally referred to as American National Standard Institute (ANSI) promulgations. Some of these standards are NFPA, IEEE, ASTM, ASM, ASCE, and numerous others. Please refer to Chapter 15 for a list of these various organizations.

Unforeseen Events

No matter how much we strive to protect the public from injury, there are events that could not be or were not recognized. A classic example of a device that was designed, and some of its consequences not recognized, is the automobile. From a historical perspective, let us look a little into the development of the automobile. Before the invention of the automobile, the main mode of transportation was the horse and the buggy. People either rode a horse or attached a horse to some form of a carriage. Contrary to what we observe in the movies or television, you cannot ride a horse at full gallop for a very long distance. What is that distance? Horse races tell us that the distance is about one mile more or less. Travel by horsepower, that is, animal horsepower and not machine horsepower, is limited to walking speeds, which is especially true if the horse is pulling a cart of some sort. In the case of a stage coach, which is considerably heavier than a small cart, a team of horses was needed in order to counteract frictional, gravitational, and aerodynamic forces. Humans who were propelled by horsepower typically did not attain significant speeds beyond that which the horse produced by walking. That speed is somewhat greater than human walking speed, which is about 4–5 mph or about 7–8 ft/s. Keep this speed in mind as we come back to it to make arguments about injuries.

When horses were the main mode of locomotion, the speeds at which folks traveled were generally restricted to less than 10 mph. With the advent of the automobile, there was a dramatic increase in the travel speed. Even the earliest vehicles were capable of producing travel speeds of 20 mph. Some even approached 40 mph. On December 18, 1898, Chasseloup-Laubat set the record of over 39 mph in an electric vehicle. By April 29, 1899, Camille Jenatzy recorded a speed over 65 mph in a dual-battery electric vehicle. On January 12, 1904, Henry Ford set the record of 91 mph with a 4-cylinder internal combustion engine–powered vehicle. By November 13, 1904, Paul Baras set the record exceeding 100 mph in a Gordon Bennett vehicle at 104.53 mph. The rest, as they say, is history.

During the first 60 years of the development of the automobile, most vehicles for public purchase were capable of producing travel speeds that were greatly in excess of what humans had evolved to produce by their power or the power of animals. The designers of vehicles during the first half of the twentieth century simply did not consider, or avoided, the danger inherent with speed. The industrial revolution along with the mechanization of processes has greatly benefited mankind but has also produced situations, machines, and devices that can maim or kill us. A lot of these developments in mechanization were and are associated with a variety of dangers.

Another example of a product with dangerous consequences is what resulted from the discovery of radiation that occurred around the turn of the nineteenth century by Wilhelm Roentgen. From the work of Roentgen, Henri Becquerel, J.J. Thompson, Ernest Rutherford, and Marie Curie on radiation and radioactivity sprung a variety of products and devices that were alleged to produce health benefits. Most of these products and devices produced ionizing radiation that was energetic enough to dislodge electrons from atoms and cause a variety of maladies in humans. These included toothpaste, creams, chocolate bars, luminescent watch faces, water jars, sexual function devices, rectal devices, and shoe fluoroscopes. In this case, science did not develop fast enough to reveal the dangers posed by radioactive devices intended to produce health benefits. The danger of these radioactive devices was indeed unforeseen. On a cruel note of fate, and a validation of Murphy's law, Marie's husband Pierre died in a carriage accident in 1906. So much for our discussion on the safety of horse-driven vehicles relative to the automobile! For the uninitiated, Murphy's law states that anything that can go wrong will go wrong. A more precise description of Murphy's law is Gumperson's law of perverse opposites that says "the contradictory of a welcome probability will assert itself whenever such an event is likely to be most frustrating." Dr. R.F. Gumperson was a physicist who noted that the work of the weather bureau was not as accurate as the Farmer's Almanac at predicting and forecasting the weather. Gumperson worked for the war services during World War II and died in 1947 by, what else, his law. One night he was walking to the left opposite traffic, wearing light clothing, obeying typical pedestrian rules when he was struck from behind by a vehicle driven by a visiting Englishman driving on the wrong side of the road.

Unfortunately, when radioactive elixirs, creams, and devices were being widely sold and distributed, the practice of plaintiff's litigation was in its infancy. Times have changed, and today there is an abundance of litigation concerning materials and devices that are deemed injurious or alleged to cause injury to humans. These may be unforeseen or may be widely recognized by a particular industry but covered up to avoid litigation. The truth of the matter is that most decisions to modify a product for safety reasons are economic in nature.

Another example of unforeseen events involves the use of AstroTurf. When this product was first developed, it was regarded as a wonder product for the sports industry. The carpet could be installed in a playing field and easily maintained—no need to cut the grass and water the field. It also provided a playing surface for indoor stadiums that were not subject to sunlight as required for proper maintenance of natural grass. However, the designers of synthetic field coverings were not aware of the different characteristics of AstroTurf with the characteristics of the shoes worn by the players. Synthetic grass simply has different frictional and directional movement characteristics than natural grass. Sports medicine specialist and trainers soon saw a variety of injuries that affected the players. Some of these injuries affected ankles, knees, and feet. Biomechanical experts were instrumental in addressing these problems, and soon the shoe designers made modifications for synthetic surfaces.

Some of us who are a little older rode bicycles as kids. Helmets for kids were not available to ride a bike. Yes, you could wear your football helmet to ride a bike, but you would have been ridiculed. Similarly, when you played little league baseball you were not equipped with a helmet as part of your gear. Helmets for motorcycles were also rare. When science revealed that the forces and accelerations produced by striking baseballs and falls from cycles on the cranial structures of humans produced profound injuries, the rules changed for the better. The realization of the effects of closed head injuries on humans has had a significant effect on the recreational, organized, amateur, and professional sports industry. We now recognize that the brain is subject to internal injury without the external symptoms that are commonly associated with such events. We can no longer say that a wild pitch was an accident that occurred and injured the batter. The wild pitch causing head injury is a predicted event according to the rules of chance, and we have developed equipment to minimize or eliminate its severity.

There are many more examples of unforeseen events that have occurred throughout modern history. Most of these were not predicted or recognized. Only when injuries became more prevalent did science become aware, studied the events, and developed solutions for their remediation. Biomechanical calculations on human tolerance play a significant role in the design, mitigation, prevention, and likelihood of injury to humans.

Faulty Equipment

Complex equipment requires complex maintenance. Most of the devices that are designed today are quite intricate from a mechanical, structural, electrical, or systems standpoint. Some of these devices are capable of producing injury to humans if they become faulty. Equipment may become faulty as a result of a lack of maintenance, a worn part, or a combination of the two. Improperly lubricated machine parts can fail and injure people. Machines that are not kept at operating temperature can overheat and fail causing a variety of injury scenarios. Extremely low temperatures can affect systems. Electrical and electronic devices have fairly strict operating temperature requirements. Operating these devices outside of their normal range can cause failure of the device or system and create the potential for injury to humans. Faultily designed equipment also has a great potential to injure.

Machine guarding is a requirement for devices that have the potential of grabbing or pinching clothing or body parts. Loose clothing around moving equipment is a great danger. Rotating equipment can grab clothing or body parts and create great injury and death. Reciprocating equipment also has great potential for creating havoc. Eye and ear protection is a must around flying debris or caustic compounds and liquids. Protective gloves are required where materials or liquids can damage the skin. In particular, work around electrical equipment that is energized has the potential for serious injury and death. Protective insulating boots, aprons, blankets, and gloves are necessary for work around energized medium and high voltage systems. The protective equipment for work in such systems, that is, the gloves and insulating materials, needs to be constantly checked for possible breakdown. Not only the safety equipment, but also the equipment that is deemed faulty has the potential for injury.

Another area where there is great potential for injury with respect to faulty equipment is in terms of fall protection. Anytime a worker is elevated above 4 ft to perform work, the worker needs to be protected from falling. The scaffolding, hoist, or ladder being used must be in proper working order. Additionally harnessing of the worker must also take place. Federal standards for fall protection of workers are very strict and all encompassing. Additionally, there are a multitude of professional standards that have been developed to protect workers in virtually all industries. Knowledge of these standards and recommended practices is an invaluable aid in assessing a particular injury to a human. These regulations help in placing limits on the possibility for injury analysis. It is imperative when making calculations with respect to injury to place upper and/or lower bounds on the severity and possibility of the injury potential.

Faulty Design

A machete is designed to be used in a particular manner. It is to be used to hack through brush away from the body of the individual user. It is used in jungles and densely wooded areas to clear paths so that people may walk through. It is not to be used to cut grass. So in this example, let us say that a homeowner uses a machete to cut grass around a bed of roses swinging the blade horizontally at ground level and cuts his bare foot, severing an artery, causes him to bleed to death. Is the machete improperly designed? Of course not, it was being used in an improper manner. Should there have been warnings on its usage? It might be a question for a jury to decide.

Today, most products have a variety of warnings on the use of equipment. From lamp cords, to coffee makers, to drills and saws, and general commercial equipment, there are a multitude of warning labels that are placed on the devices and in the instruction manuals. These warning signs and messages are intended to specify what the device or equipment is designed to do and how it must be used. However, sometimes a device is used as intended in a proper manner, but it still may cause injury. One of the most high-profile devices that, it could be argued, was improperly designed is the airbag in vehicles. When originally designed, airbags in vehicles deployed very rapidly and violently. As a consequence, some smaller motorists, especially small women and children, were prone to be injured by airbag deployment as a result of a crash. After a multitude of injuries and deaths occurred, engineers in the automotive industry redesigned the airbag systems so that they were less energetic and smaller in size when deployed. This redesign of airbag systems has greatly reduced injuries produced during deployment. Some bruising and abrasions still occur during deployment, but injury has been greatly diminished. It is, after all, better to be bruised rather than seriously injured or dead.

Two of the most famous faultily designed automobiles were the Chevrolet Corvair in the 1960s and the Ford Pinto in the 1970s. The Corvair was a radical new design with a rear-mounted engine. This vehicle had many attributes such as roominess and good traction in wet, snowy conditions because the center of mass was very close to the rear axle. However, the location of the center of mass made the vehicle rather unstable so that the famed consumer advocate, Ralph Nader, coined the phrase "unsafe at any speed." The Pinto had a rear-mounted fuel tank that was subject to bursting upon a rear impact. This effect produced many fires that injured and killed people.

Another more recent example of an alleged faulty design involves the accelerator of some late model Toyota vehicles. Most of us remember older model vehicles that had mechanical linkages from the accelerator pedal to the carburetor that metered the amount of fuel and air that was required for combustion. The more air and fuel delivered to the cylinders, the faster the engine turned and the more speed that was achieved. Any accelerator failure could only be produced by a failure of the mechanical linkage. Today, most vehicles are designed with fuel injection systems that deliver the fuel to each cylinder from a common rail. Each cylinder is metered by an injector that is electronically controlled by the accelerator pedal. Actually, the pedal itself is not mechanically connected to the injection system. The connection is electronic and regulated by the computer system of the vehicle for optimum performance and minimal pollution. There are two methods of failure for these systems. One is that the accelerator pedal could get stuck by an object such as a floor mat. The other method of failure could be electronic. Any electronic failure would be recorded by the vehicle's system computer. The stuck accelerator claims were analyzed in most instances and were attributed to problems associated with the floor mats. Under certain circumstances, the floor mats would become lodged in a manner that depressed the accelerator pedal. In one highly publicized case in California, a driver claimed that the accelerator stuck and he was not able to stop the vehicle for numerous miles even though police were following him after he reported the incident on his cell phone to 911. Finally, he was able to bring his vehicle under control. The case was investigated by the National Transportation Safety Board. The mistake that the claimant made was that he was not aware that the vehicle's computer would record the events leading up to the stoppage of the vehicle. The computer revealed that the accelerator and then the brake were depressed repeatedly over 200 times during the course of the reported malfunction. The claimant and his attorney simply went away.

Most brake systems are more robust than vehicle accelerator systems. Any kid has tested this fact when he decides to peel out and lay some rubber. The common method is to depress the brake, rev the engine through the accelerator, and then release the brake. Drag racers do this maneuver each time they come up to the line. What happens? The brakes keep the vehicle from moving. So, the brake system of a vehicle is sufficient to overcome the action of the accelerator. Most stuck accelerator claims are actually a result of the driver depressing the accelerator instead of the brake. We have actually worked such a case, and when the computer system was interrogated, it was found that the accelerator was depressed when the brake should have been depressed. It appears that older drivers are susceptible to this confusion, especially when backing the vehicle. A likely occurrence is that the pedals are confused while performing the backup maneuver because the body position is taken out of its normal situation.

Predictable Events

By predictable events we do not mean that in all cases, the event is predicted prior to its occurrence. Actually, most predictable events are seen in hindsight. If someone had just looked, they may have been able to tell what may or could happen. Sometimes the outcome is so obvious as when a worker is elevated without fall protection. He is on a steep roof, loses his footing, falls, and is injured. The extent of the injuries is directly related to the distance that the worker falls. The farther the worker falls, the greater the speed at

Table 3.1   Fall Speed and Distance

Speed v (ft/s)	Distance D (ft)	Speed v (mph)
8.0	1	5.47
11.3	2	7.74
13.9	3	9.48
16.0	4	10.95
17.9	5	12.24
19.7	6	13.41
21.2	7	15.48

which he strikes the ground. We can do a basic calculation for the speed of a falling object, such as a worker or a person jumping from a height as will be explained soon at the end of the chapter.

Table 3.1 and Equation 3.1 compute the speed in feet per second and miles per hour for fall distances from 1 to 7 ft and will explain why the federal standard requires fall protection above 4 ft in elevation. It also explains the onset of injury in other sections of the book relative to speed change experienced by folks when they fall or generally collide with objects or are struck by different bodies or surfaces. In mechanics we refer to any object as a body. A body has mass and occupies space. Bodies are not necessarily human bodies. We make every attempt throughout the book to differentiate between the two in order to avoid confusion.

We see from Table 3.1 that at a fall distance of 4 ft the speed at impact of the body is approximately 11 mph. A multitude of tests have shown that when humans strike objects or are struck by objects at approximately 12 mph, injuries begin to occur. This is the widely recognized scientific threshold speed for injury. At striking speeds above 17 mph corresponding roughly to falls of 8 ft, there is a likelihood of injury approaching 100%.

Some explanation is necessary at this point. This is not to say that a trained athlete cannot jump down from a height of 8 ft although the average person has a great potential of at least twisting an ankle, injuring a knee, or suffering some other nonincapacitating injury. The trained athlete knows how to jump and absorb the shock through his muscles, tendons, ligaments, and bones. If the athlete simply falls, he may suffer some damage. The average person making the jump may well survive without injury, but the chances are not on his side, especially if he falls. Older individuals are, of course, more susceptible to injury. The chapters that follow on the strength of biological materials clearly outline the correlation between aging and the degradation of the strength of our biological systems.

The airbag example we gave in the previous section is also a class of events that could have been predicted had sufficient testing been performed. In many instances, monetary constraints limit the predictability of an outcome.

The space shuttle Challenger is an example of a predictable event. On January 28, 1986, at an ambient temperature of approximately 28°F, the orbiter catastrophically broke apart a little more than 1 min after liftoff. A right side o-ring failed and allowed gasses from the solid rocket booster to affect the surrounding structure, causing the right side of the vehicle to separate. The combination of the orbiter's speed and the aerodynamic forces that developed broke apart the shuttle. The previous low temperature for a launch of an orbiter had been at 53°F. As early as 1977, tests by Morton Thiokol, the contractor for the solid rocket boosters, determined that a phenomenon referred to as "joint rotation" allowed combustion gasses to erode the o-rings. The o-rings were utilized to field join sections of the solid rocket boosters. Engineers from the Marshall Space Flight Center in Houston, Texas on several occasions suggested that the design of the field joints was unacceptable. This information was not relayed to Morton Thiokol so that the design of the field joints was accepted. By the second orbiter flight, erosion of the o-rings was present, but Houston did not report the problem to NASA management. In 1985, Morton Thiokol and the Marshall Space Flight Center realized that the o-ring problem was potentially catastrophic, and a redesign of the joints for the solid rocket boosters was initiated. Shuttle flights were not stopped during this process, and the problem was deemed an acceptable risk.

On the day of the launch the weather was cold, and there was considerable ice that had accumulated on the orbiter. Concerns were raised about the cold temperatures, and managers at Morton Thiokol supported recommendations from its engineering staff to postpone the launching of Challenger. NASA opposed the recommendation to delay the launch pressuring Morton Thiokol to reverse its recommendation. Morton Thiokol was a subcontractor for the shuttle program. The main contractor, Rockwell International, was also concerned about the large amount of ice that had accumulated at the launch pad. Liftoff of Challenger occurred at 11:38 a.m. eastern standard time.

In retrospect, the chain of events that occurred in the Challenger disaster is clear and is a classic example of predictable events that can lead to injury or death. Many of these events can lead to civil or criminal prosecutions, which is true whether the loss is purely financial or whether it includes some form of injury. In this book, of course, we are interested in injuries that may occur and then lead to some form of litigation.

Litigation from a civil standpoint occurs when a party is injured, and blame can be assigned to another party. As we have outlined in this chapter, the injury can result from a variety of sources. Consequently, the maligned party may proceed with litigation in order to be compensated for his injuries. In some instances, injuries may be claimed by an event for which the analysis of the injury does not support the allegations. This type of claim is prevalent in low-speed collisions that may occur or may be staged. In some instances, the injuries may be preexisting or may be exaggerated in order to benefit financially.

From a criminal perspective, a biomechanical engineer may be employed to assist in the determination of an injury or death. This type of assignment may come from the prosecution or the defense. The case could involve equipment that is suspected of causing the loss. Usually, this type of case would involve a variety of experts so that the biomechanical engineer would only be one cog in the wheel of justice.

As an example, a basic computation may involve determining the speed that an individual achieves by jumping or falling from various heights ranging from 1 to 7 ft to the point where he lands. We are all aware that active children have no fear so that we often see them jumping from various objects of varying heights. Neglecting air resistance, the equation for a falling object is given as

$$v = \sqrt{2gD} \tag{3.1}$$

where
    g is the acceleration due to gravity (ft/s²)
    D is the distance the object falls (ft)

Table 3.1 shows the speed in feet per second and miles per hour achieved by falling through distances from 1 to 7 ft. This table is critical in that it explains why the injury demarcation is generally considered to be about 12 mph in speed change.

Types of Injuries

4

Injuries to humans can occur to any part of the body. However, we concentrate on injuries that are caused by some event as described in Chapter 3. Injuries due to the incidents described exclude disease and naturally occurring events such as stroke, heart attack, cancer, and other ailments that cannot be correlated to a physically traumatic source. The types of injuries we are concerned with are a result of some form of a physical activity such as a sport, trauma caused by a fall or a blunt or sharp instrument, trauma resulting from a vehicular collision, or physical force exerted by another individual. The trauma may also be caused by a device or machine that is not well guarded. The trauma may also be caused by some form of a criminal activity. In simple terms, mechanical action is necessary to produce the type of injuries we will deal with.

We choose to divide the types of injuries to the following categories and discuss each category in greater detail. We begin with the head. The head is separated into the following subcategories: closed head injuries, injuries to the cranium, injuries to the face and jaw bones, and finally injuries to the teeth. Note that we will not cover injuries to the eye because of its very delicate nature. Injuries to the neck are divided into three categories: the cervical spine, the supporting muscles, and the vessels and soft tissues, including the Adam's apple. The more proper term for the Adam's apple is the laryngeal prominence, which is more defined in the male than in the female. The thorax or chest cavity section of the body is comprised of the following categories: the thoracic spine, the ribs, shoulder bones, shoulder joints, muscles, tendons, ligaments, cartilage, soft tissues, and internal organs.

The hip girdle is comprised of bones, the lumbar spine, the tail bone, the hip joints, tendons, ligaments, cartilage, muscles, and abdominal cavity. The upper extremities consist of bones, elbow and wrist joints, hand and finger joints, tendons, ligaments, cartilage, and muscles. The lower extremities consist of bones, knee and ankle joints, toe joints, tendons, ligaments, cartilage, and muscles.

For each of these sections and subsections, we detail the most common type of injury that occurs along with the main function of each body part. Chapter 7 details the basic elements of anatomy that define the various body parts that can be injured. Chapter 6 discusses some basic biomechanical terminology that further identifies the location and exact form of the ailment and injury.

The Head

We begin with the interior contents of the head and limit this discussion to the brain. Catastrophic injuries to the head would include a penetration into the skull. However, we separate those injuries to the closed head and all others. Injuries to the head that do not penetrate the bony structures are also very damaging. Generally, these are referred to as concussions. These closed head injuries where no penetration of the skull occurs are a result of two or three separate impacts. The first impact might occur when a vehicle is involved in a collision that allows the head to strike an interior surface of the vehicle. This is the second collision. The third collision would occur when the brain strikes the interior surface of the skull. This type of injury may also be produced when two football players collide producing the first collision, then the respective skulls would strike the interior surfaces of the helmet producing the second collision. Finally, the third collision is between the dura and the brain itself. The third collision with the brain may involve the front, the rear, or the side of the brain. It may also involve rotation as the brain moves at a different rate than the skull.

Most closed head injuries occur as a result of car accidents, sports injuries, falls, and acts of violence. A two-impact head injury can result from a blow to the head with a fist or an object such as a baseball bat or a policeman's Billy club. A one-impact head injury can result from shaken baby syndrome where an adult violently shakes an infant producing severe accelerations and decelerations of the brain. Closed head injuries account for approximately three-fourths of all head injuries and about one-third of those are produced by falls.

Figure 4.1 shows a cross section of the skull. The membrane that surrounds the brain consists of three layers with the dura being the outermost layer.

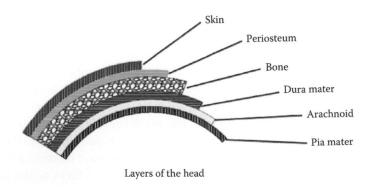

Skin
Periosteum
Bone
Dura mater
Arachnoid
Pia mater

Layers of the head

Figure 4.1 Cross section of the skull.

The inner most layer is the pia mater, followed by the arachnoid mater, and then the dura mater. Closed head injuries vary in intensity from the most severe such as diffuse axonal injuries resulting from car accidents to the least, which are referred to as concussions. In between these extremes, there are cerebral contusions in which the brain is bruised and the lesser intracranial hematoma resulting in blood vessels rupture. These may be epidural (between the brain and the skull) or subdural (around the brain).

Closed head injuries are classified according to the Glasgow Coma Scale. This scale is divided into 15 points. The most severe between one and three is reserved for vegetative states of the injured individuals. Severe brain injuries vary from 4 to 8, and the milder injuries, ranging from 9 to 15, account for the 15 points. Another method of quantifying closed head injury is the Head Injury Criterion (HIC) generally used in automobile and sports related cases. This criterion, which was developed from scientific studies by a variety of researchers, compares the accelerations to the time element in a head collision. A greater acceleration with respect to a shorter time element results in a higher criterion and consequently a higher propensity for injury. Pressures associated with the accelerations produced are plotted with respect to a HIC number to signify the onset of injury. The original HIC number for the onset of injury was set at 1000 but has been lowered in Federal Standards to 700. The Head Injury Criterion is mathematically defined in Chapter 7.

Severe injuries to the cranium involve fractures to the bony covering of the head as well as fractures to the facial bones and the jaw bones. Some of these fractures can also occur to the teeth. One ailment sometimes attributed to traumatic injury is the temporomandibular joint (TMJ) disorder. The TMJ connects the lower jaw known as the mandible to the temporal bone of the skull. This joint consists of muscles, ligaments, discs, and bones that allow for complex movement required for talking, eating, and other associated movements. Disorders of this joint may be due to a variety of causes such as clenching of teeth, dislocation, arthritis, stress, or a blow. Some people attribute TMJ to low-impact rear-end collisions. However, there is no scientific validity to these causes. Figure 4.2 shows a side view of the cranium with the major bones identified.

The Neck

The neck can be injured in a variety of ways. The most common neck injuries involve the cervical spine. One noninjurious condition of the neck is cervical spondylosis better known as arthritis of the neck. This condition generally results from the aging process and a degeneration of the joints of the neck. Often this condition is alleged to result from low-impact rear-end collisions involving older individuals. This condition rarely results in crippling effects,

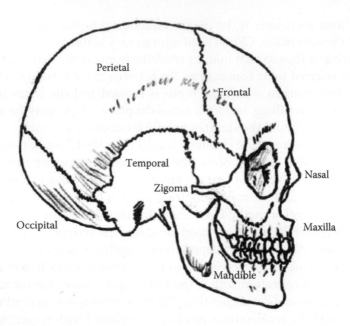

Figure 4.2 Bones of the cranium.

and it is estimated that approximately 85% of individuals over the age of 60 suffer from the condition. The other elements of the neck that are generally injured include the supporting muscles, the vessels, tendons, ligaments, and the soft tissues. The main supporting muscles are the trapezius muscles and the sternocleidomastoid muscles that help support and flex the neck and head. These muscles are generally injured when whiplash occurs in an event such as rear-end collision in an automobile. Another type of impact that may involve these muscles results from two football players coming into contact while a tackling maneuver is undertaken. Figure 4.3 shows the main neck muscles. There are many other muscles in the neck that may also be injured along with the trapezius and the sternocleidomastoid. The main vessels that are affected are the trachea, known as the wind-pipe that extends from the larynx (better known as the voice box) branching to the bronchi leading to the lungs, the esophagus which is the tube extending from the throat (pharynx) to the stomach, and the carotid arteries and jugular veins (see Figures 4.4 and 4.5). Recall that arteries carry oxygenated blood from the heart and veins return the blood to be reoxygenated in the lungs. On a general note, asphyxia means that the pulse has been stopped and generally refers to a multiple set of conditions that result in inadequate processing of oxygen. Basically, air is prevented from entering the lungs and/or blood to the brain is cut off. One or a combination of these conditions will cause death. During strangulation, excess pressure to the vagus nerve can also cause signals to the brain to stop the heart from beating.

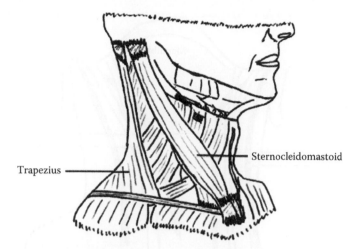

Figure 4.3 Main neck muscles.

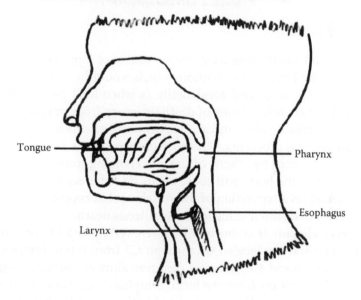

Figure 4.4 Cross section of the neck.

There are a total of 12 pairs of nerves that extend from the medulla to the rest of the body through the structures of the neck. The vagus nerve is actually the 10th cranial nerve that exits in pairs from the medulla and extends through the jugular foramen and passes between the internal carotid artery and the internal jugular vein. The vagus nerve acts to lower the heart rate, which is said to control the parasympathetic innervation of the heart. The medulla oblongata is the lower half of the brain stem that is continuous with the spinal cord. The medulla involves involuntary functions of heart rate, breathing, and blood pressure.

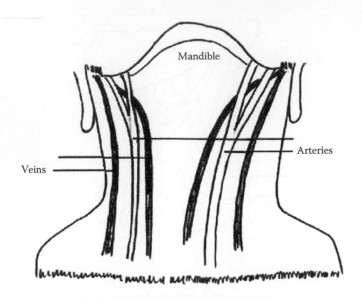

Figure 4.5 Blood vessels.

Choking and smothering are common causes of asphyxiation. Choking can be caused by human intervention, which would be classified as strangulation, or it may be caused accidentally as when food gets lodged in the windpipe. There are other forms of death or injury from oxygen deficiency. Some of these include carbon monoxide poisoning, chemical exposure, sleep apnea, hanging, and drowning. The cause of these events may be accidental, self-induced, produced by others, or suicide. Hanging from a short distance or by suspending the body will cause asphyxia, whereas hanging as in an execution produces a separation of the vertebrae that ruptures blood vessels and tears the spinal cord leading to instantaneous death.

The cervical spine is comprised of seven vertebrae and corresponding discs. The vertebrae are labeled C1 through C7 from top to bottom. Please refer to Figures 4.6 and 4.7. The most common ailment resulting from a collision in a vehicle results from the term "whiplash" relative to such an incident. These events normally are associated with rear-end collisions but may be produced in other configurations of a vehicular collision. These injuries can result from sufficient speed changes produced by the relative motion of the head with respect to the thorax, which is well supported by the seat back. Head movement is restricted by the headrest so that spinal injuries to the cervical portion of the spine require sufficient accelerations in order to be produced. Other portions of the book deal in detail with such injuries. Commonly, the complaint involves bulging discs, which could have been produced in the collision. X-rays and MRIs are required to medically assess these injuries, which can then be correlated to biomechanical calculations and the medical history for the patient. From an evidentiary standpoint,

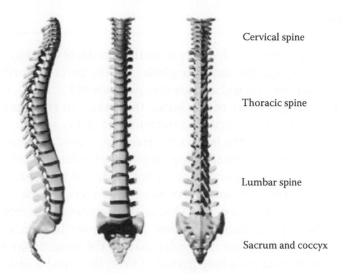

Cervical spine

Thoracic spine

Lumbar spine

Sacrum and coccyx

Figure 4.6 Spine.

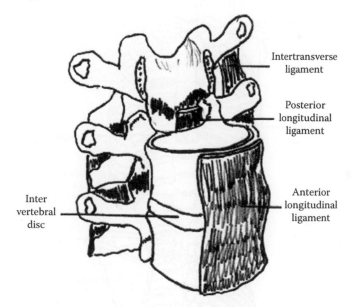

Intertransverse ligament

Posterior longitudinal ligament

Anterior longitudinal ligament

Inter vertebral disc

Figure 4.7 Vertebra.

simply relating a whiplash injury to a treating chiropractor is not considered scientific. Such claims can only be assessed through medical diagnostic tests and substantiated through biomechanical calculations. Biomechanical calculations are an important scientifically valid avenue to determine the possibility of such injuries occurring. Without these calculations, it is scientifically insufficient to determine causality without the combination of imaging medical techniques and biomechanics.

The Thorax

The thorax is the portion of the body from the shoulders down to the waist. It includes the rib cage, the shoulder girdle, and the thoracic portion of the spine. It excludes the cervical and lumbar portions of the spine. Figure 4.8 shows front and rear views of the bones that make up the thorax. These include the scapula, the clavicle, the sternum, the spine, and the ribs. Note that not all the ribs are attached to the sternum. These are referred to as floating ribs. Ligaments connect bones to other bones and are comprised of fibrous tissue. The connection forms a joint. It should be pointed out that ligaments do not connect muscles to bones. Tendons are the connective tissue between muscles and bones. One of the features of ligaments is that they are viscoelastic, meaning that they shrink under tension and return to their original shape when not under load. If the tendons are stretched beyond a certain point for a long period of time, they tend to lose the ability to return to their original shape. When this occurs, the joint being

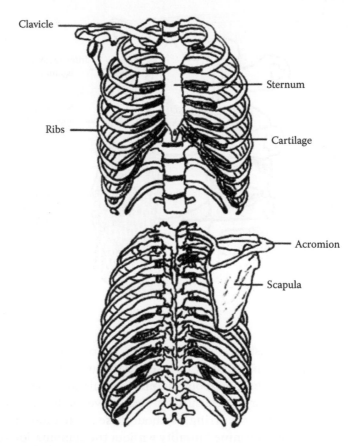

Figure 4.8 Thorax bones.

held by the ligament will be weakened. As an example, when a shoulder is dislocated and not set soon after, the shoulder joint may weaken producing a propensity for future dislocations. Dislocations are most common to the shoulder, but they can occur to most parts of the body. Some other examples of dislocations are the knees, hips, toes, or fingers or the elbows. Dislocation is referred to as luxation in medical terms. If the dislocation is partial it is called subluxation. This is a common condition experienced by athletes. Some people have very elastic ligaments and are commonly referred to as double jointed. The actual term to describe this condition is hypermobility. Athletes of all types tend to exercise their joints to stretch the ligaments and have a greater range of motion and therefore a greater propensity for luxation in combination with strenuous activity. Trauma in a variety of forms can cause luxation or subluxation.

The most common ligaments that aid in movement are called articular ligaments. In addition, there are other less numerous ligaments. These include capsular ligaments, which are reinforcements to the main ligament and are contained within the capsular sheath of the articular joint.

Generally, it is very difficult to break a ligament because of its viscoelasticity. However, it does happen and often requires a surgical procedure to correct the problem. If the problem is not corrected, the joint will most likely become unstable. An unstable joint will not rotate and move as designed, which will eventually wear in a nonuniform manner on the surfaces of the cartilage. The uneven wear of the cartilage that proceeds to the point where bone wears on bone is the condition known as osteoarthritis. Many people suffering from this condition may be involved in an accident and attempt to blame their condition on the accident. This allegation may be true whether they fall, are involved in a vehicular collision, or are involved in an altercation with another individual. Figure 4.9 shows the main muscles of the thorax. These diagrams only reveal the most common joints, ligaments, and tendons of the thorax. These body features are also the ones that are most prone to injury. Figures 4.10 and 4.11 show the shoulder muscles and joint.

Cartilage is a flexible connective tissue usually found at the ends of bones. Some typical examples of cartilage are found in the ribs, nose, intervertebral discs, knees, elbows, and shoulders. Cartilage does not contain blood vessels. Cartilage that is generally found at the wear surfaces between bones is called articular cartilage. If the wear surfaces of the joint wear away the cartilage, the condition is osteoarthritis, which is not considered a disease. Sometimes, the cartilage can be subjected to a rupture produced by trauma. This type of injury is common to the knee and to the spinal discs. A herniation of the discs is normally created by a nonsymmetrical compression of the discs between the vertebra causing the ruptured disc to encroach upon the nerves of the spine, which then produces pain. Figure 4.7 shows a vertebra, a section

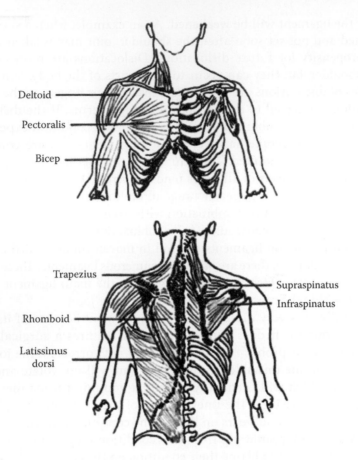

Figure 4.9 Thorax muscles.

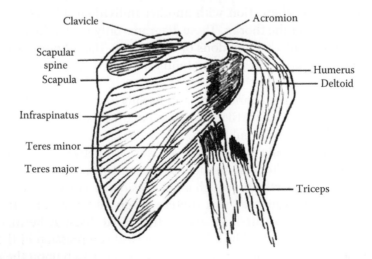

Figure 4.10 Shoulder muscles.

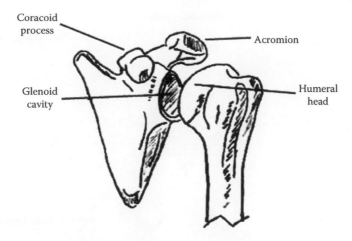

Coracoid process

Acromion

Glenoid cavity

Humeral head

Figure 4.11 Shoulder joint.

of the spine, and the entire spine is detailed in Figure 4.6. Two common terms associated with spinal problems are radiculitis and neuritis. Radiculitis implies a change or disturbance to the nerve function and neuritis means that there exists an inflammation of the nerves. These conditions are generally produced when a bulging disc impinges on the nerves as they pass through the spinal canals. A term associated with an abnormal narrowing of a canal or tubular structure is stenosis. Stenosis then refers to narrowing of blood vessels, bronchial tubes, carpal tunnels, and digestive organs and is often encountered when medical diagnosis is reviewed. Spinal stenosis refers to the narrowing of the passages in the spine, which may occur in any region. The main causes of stenosis in the spine are aging, arthritis, heredity, trauma, and tumors. A diagnosis is the purview of the medical expert and not the biomechanical engineer.

The Hip Girdle

The main components of the hip girdle include the lower portion of the lumbar spine, the sacrum and coccyx, and the pelvis in general. Figure 4.12 shows the pelvis. Figure 4.13 depicts the hip joint in greater detail.

Injuries to the pelvis include some form of a break. The most common reasons for a broken pelvis are falls, especially in the elderly, running, in which the leg places undue stress near the hip joint, and car accidents, in which the impact may be direct to the pelvis or transferred through the legs. Breaks to the pelvis or dislocations of the hip joint commonly occur when one vehicle strikes another vehicle or some form of a rigid barrier, and the driver attempts to stop the vehicle by applying the brake pedal. The force is transferred through the leg to the hip joint, which causes the break near the

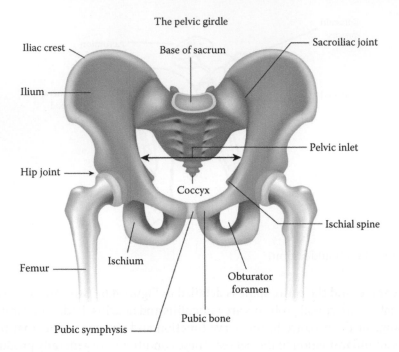

Figure 4.12 Pelvis.

lower portion of the pelvis at the acetabulum, pubis, or ischium. If the pelvis does not break, it may well be dislocated.

A classic case of a dislocation produced while running involves the famous athlete Vincent Bo Jackson. Jackson was a two-sport professional athlete competing in football and baseball. During the 1990 playoff game against the Cincinnati Bengals, he was tackled from behind by Kevin Walker reportedly dislocating his hip joint. Jackson claims that he was able to pop the joint back in place. The injury caused a deterioration of the head of the femur as a result of decreased blood supply known as avascular necrosis. This condition produced a deterioration of the femoral head that required a hip replacement and eventually ended his career in both sports. A review of the film and the tackle revealed that his leg was trapped in the tackle as Jackson was accelerating forward with the other leg. The extra weight of the tackler and Jackson's dynamic strength led to the hip dislocation. Figure 4.14 shows the main muscles that form the hip girdle.

Lower Extremities

The main components of the lower extremities include the upper leg, the knee, the shins, the ankle, and the foot. A word of caution is necessary at this point with reference to leg and arm as will be apparent in Chapter 6.

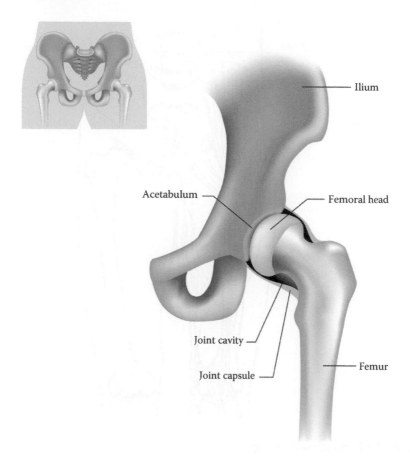

Figure 4.13 Hip joint.

The knee is shown in Figure 4.15 showing the ligaments, cartilage, meniscus, the femur, tibia, and fibula. This view shows the rear of the knee. Common injuries to the knee include torn meniscus, ligament damage, and cartilage damage. Cartilage deterioration also results from age and wear and tear. The anterior cruciate ligament (ACL) is commonly injured in sports. The other ligaments to the medial, lateral, and posterior may also be injured.

Figure 4.16 shows the femur. Most serious injuries to the femur include breaks near the head, the shaft, or near the condyles. The joint wear surfaces of the head and the condyles may be affected through injury, through the aging process, or through disease. A proper diagnosis by a medical expert is required to make that assessment. Figure 4.17 shows the bones of the lower leg. As with the upper leg, the shafts of the fibula and tibia may be broken as a result of insult. The condyle and malleolus can also be affected including the ligaments, tendons, and wear surfaces. The muscles of the lower leg are not shown in this chapter but are detailed in the anatomical portion of the book. A significant injury to the lower leg involves the rupture of the

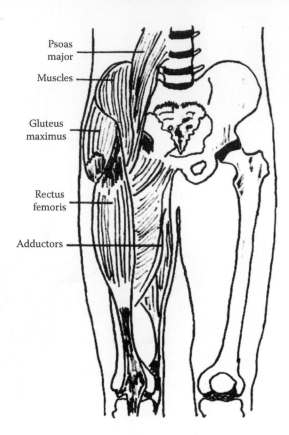

Figure 4.14 Muscles of the hip.

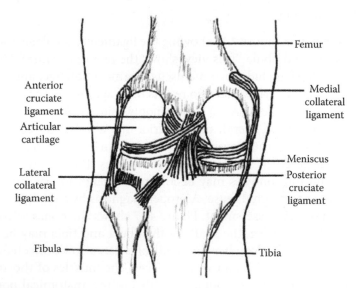

Figure 4.15 Knee.

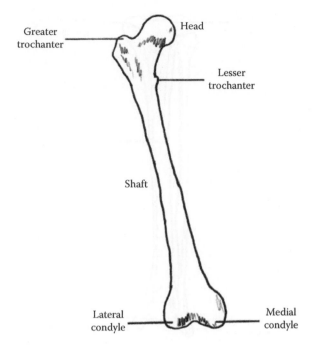

Figure 4.16 Femur.

Achilles tendon, which is the strongest and thickest tendon in the human body. This tendon attaches the gastrocnemius, soleus, and plantaris to the calcaneus. The main forms of injury include sudden flexion of the ankle through physical activity, trauma, atrophy, and age. The gastrocnemius is the muscle at the back of the leg commonly referred to as the calf muscle. The soleus muscle is the other part of the rear leg muscles of the calf and is located in the superficial compartment of the posterior of the leg. The plantaris muscle is another thin structure in the calf area and is not present in all humans. Approximately 90%–93% of humans have a plantaris muscle. The calcaneus is the heel bone. So, the sum of the calf muscles attach to the heel bone through the Achilles tendon.

Figure 4.18 shows the leg bones and Figure 4.19 shows the ankle and foot.

Upper Extremities

The upper extremities are detailed in Figures 4.20 through 4.22. Naturally they are generally referred to as the arms, which is a medical misnomer. As we will see in Chapter 6, it is not anatomically correct to refer to the upper extremities as arms and the lower extremities as legs. The bone of the upper arm is the humerus, which joins the lower bones through the elbow joint.

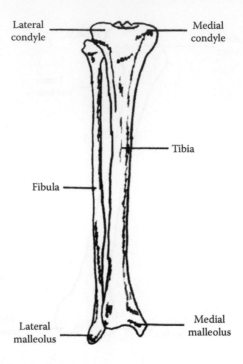

Figure 4.17 Lower leg bones.

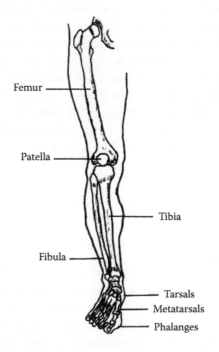

Figure 4.18 Leg bones.

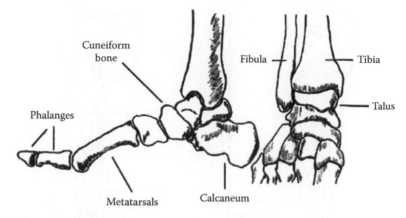

Figure 4.19 Ankle and foot.

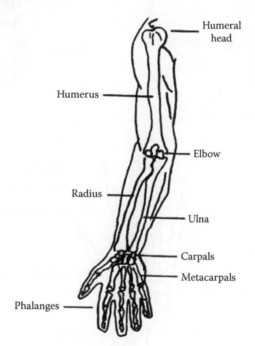

Figure 4.20 Arm bones.

The head of the humerus attaches to the shoulder through the humeral head. The motion surfaces are lined with cartilage and tendons. Tendons attach bones and muscles to each other. The lower arm bones are the ulna and the radius. The wrist attaches the lower arm bones to the hands through carpals, metacarpals, and phalanges. The muscles include the deltoid, spinatus, biceps, triceps, palmaris longus, extensor digitorum, and related tendons. As with the lower extremities, any of the associated structural elements can be injured though stress or physical insult.

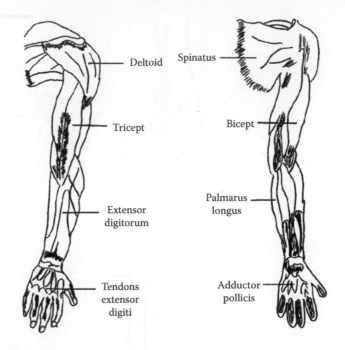

Figure 4.21 Arm muscles.

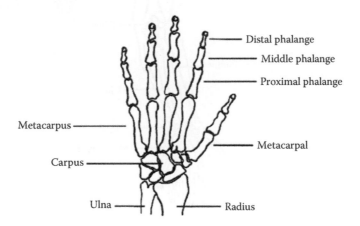

Figure 4.22 Hand bones.

A common rupture of the elbow joint occurs during the sport of arm wrestling. Arm wrestling puts tremendous stresses on the biceps, triceps, and elbow. A common injury to the elbow from arm wrestling is tendonitis. Tendonitis refers to the inflammation and irritation of the tendons. The elbow joint may also break during this sport. Tennis elbow or golf elbow type injuries are common in arm wrestling. Through overuse, the extensor muscles originating from the lateral epicondylar area of the distal humerus can cause lateral epicondylitis better known as tennis elbow. The most severe injury occurs when the humerus breaks below the midpoint. Breaks can also occur to the lower bones of the arm. Generally, trauma of any sort is responsible for injuries to the upper extremities.

Another common injury to the upper extremities is a scaphoid fracture of the wrist. This injury is normally associated with falls in which the person stretches the hand to catch himself. The wrist is comprised of eight carpal bones that connect the radius and ulna bones to the metacarpals. The scaphoid bone is the small carpal that attaches radially to the thumb. A more serious injury to the scaphoid bone occurs when blood supply is lost to the distal end of the fracture so that avascular necrosis is produced. If the fracture tears the artery, the blood supply is lost and a portion of the bone dies. Necrosis means death, and avascular means loss of blood supply.

With distinct purpose we chose not to include the smaller muscles, tendons, and ligaments of the hands and feet. That is not to say that these structures are not vitally important and critical for movement and function of the human body. These structures are prone to a plethora of injuries as with any other part of the body. In the case of a physical assault on a person by another, the feet and hands are certainly subjected to trauma and injury. In many instances, they are injured and referenced as defensive wounds, especially to the hands. In general trauma, the relative size and weakness of these structures reveal that their injuries are secondary to larger structures. Consequently, we feel no distinct need to emphasize them in terms of biomechanics.

The Need for Analysis

5

There are a variety of reasons why there is a need to perform biomechanical calculations and analyses in forensic engineering. These include the protection of life and safety, the protection of equipment, to validate testing of human injury, to analyze human tolerance levels, to correlate computations with injury potential, to validate or dispute injuries to humans, to design safer equipment in order to better protect people, and to design safer machines or cars. We expand on each of these topics in detail.

Protect Life and Safety

As it has been emphasized throughout this book, the perspective from which it is written is engineering. The main charges in engineering are to protect life and to ensure the safety of humans. The general population thinks that engineering is simply involved with the design and construction of a variety of equipment, machines, and structures. While this is basically true, the responsibility is to perform these actions so that there is a significant reduction in the potential for injury of the human animal. In order to ensure safety, a variety of calculations must be made. Let us look at a few examples. We can take these examples from the simple design of a structure such as a building that will be occupied by the general population. A multitude of the elements of a building have a potential to injure people. For example, glass in doors and windows must be shatterproof and significantly resilient because people will normally bump against these features. The risk of lacerations from broken glass can maim, disfigure, and kill humans. Another example deals with hallways and doors, which are under the category of means of egress or more commonly referred to as exits or escape routes. They are necessary in case of an emergency such as a fire, a collapse of a building, or a crazed individual perpetrating havoc. Building codes require certain dimensions for the exit ways and maximum force to operate the doors. This force is, of course, produced by humans so that the strength of normal humans needs to be considered. This strength is analyzed in terms of the capability of the normal individual, which is a function of their bones, ligaments, tendons, and muscles as well as their respective joints.

Another need for analysis in order to protect life and safety deals with the common machines that we use such as the automobile. Crash tests are conducted on vehicles and human or dummy subjects. Equipment that records the forces and accelerations on the subjects is used to ensure that the person or dummy is capable of withstanding the crash with minimal or negligible injury. Once the forces and accelerations are measured ensuring that humans will not be injured, a lower limit is placed on the potential for injury. Then, when an event occurs that has the potential for injury, calculations can be made to assess the risk.

Protect Equipment

At first glance this topic seems to be somewhat ludicrous. Actually, it is not and is somewhat in consort with the first topic on the protection of life and safety. When equipment is designed to perform a particular function or a set of functions, it is often integrally tied to other equipment. Accordingly, it is necessary to determine the forces that the equipment is capable of producing. Consider two very different types of equipment, a computer and a bulldozer. There is no argument that the bulldozer can injure or kill humans or critically damage other types of equipment, which may result in injuries to humans. But what about the computer? Can it also injure humans or cause other equipment to malfunction and damage other types of equipment and consequently humans? Of course, it can. Most of the systems in a computer are of low voltage and current. Generally, these components operate at voltages less than 5 V and at currents in the milliampere range or even less. The power of these components is the product of the current and the voltage, which is also in the milli (milli means one-thousandth) or at best a few watts. The energy developed is the product of the wattage and the time element that the system exposes an individual to. Contact with the system is therefore negligible. As a very general rule, approximately 500 W of electrical power is required to injure humans. Electrical energy can injure in two main ways, it can produce a shock hazard that may electrocute the individual or it can cause a reaction that produces physical harm to the individual. Even if the individual is not electrocuted, significant internal damage to the body can occur. However, the power supply for the computer is generally the standard electrical system that is 110–120 V and protected by a circuit breaker at 15 or 20 A. The wattage available is then between 1650 and 2400 W or between three and roughly five times the wattage required to injure or kill an individual. In fact, more people are shocked or electrocuted with these low voltages than with higher voltages. Keep in mind that electrocution is associated with the loss of life, whereas shock refers to injury that does not result in death. On a side note we should explain what is meant by low voltage. According to the National Electrical Code, low voltage is defined as that which is below 600 V. Medium voltage generally refers to voltages between 600 and 35,000 V. High voltage distribution systems are above the 35,000 level.

Let us see why this amount of voltage and current is very dangerous to humans. We can do a simple analogy to analyze the situation. Consider 2400 W contacting an individual for approximately one-fifth of a second. This is approximately 500 W-s or 500 J of energy. One joule is 0.7376 ft-lb so that 500 J is approximately a little more than 350-ft-lb. Is that a lot of energy? Of course, it is. That is equivalent to being struck by a 350 lb hammer with a 1 ft handle or a 175 lb hammer with a 2 ft handle.

Another interesting analogy arises when we compare wattage to horsepower. One horsepower is equal to 745.7 W or 550 ft-lb/s. So in this analogy we would ask a daredevil to enter the ring with a horse in order to see who would win the hypothetical contest of strength. A horse can produce one horse power for extended periods of time but a man cannot. The strongest men in the world can produce near 1000-ft-lb for only a few seconds at which time they are physically spent. We think the horse would win the combat in the ring. Another way of looking at it is that a horse can carry a rider for hours, but a man cannot carry a horse at all. Of course, this is mainly due to the robustness and size differential between the two. It is like comparing a wreck between a semitruck and a passenger car. The car always loses.

Today's technology allows computers to control most of the mechanical processes in everyday living and in industry. When computers malfunction, they can create havoc and produce a variety of damage to other equipment or to humans. These low energy devices control relays and other interface components that in turn control higher energy devices through direct data links via cabling or though electromagnetic radiation in the form of radio waves.

Validate Testing

What exactly do we need to do to validate testing and how is this accomplished? There is a wealth of information on the strength of biological materials. Most of this testing has been performed on cadavers and their structures. Bone is an example of human or animal testing that has been performed in order to determine the characteristics of this material. Bone has been tested both wet and dry. The reason for both types of tests is to look at the characteristics of the material in both environments. Wet bone is, of course, more consistent with the bone in a living human or animal. Other biological materials have been similarly tested. Tests on biological materials obtained from cadavers yield information of the material itself and give great insight into its behavior. However, injuries do not occur solely to the particular material. The injury to a particular biological structure must also involve surrounding structures that tend to suppress or mediate the injury. In some cases, these structures may exacerbate the injury. One example of this effect was pointed out in the case of Bo Jackson where damage to a blood vessel produced a more severe injury.

For example, consider a knee whose anterior cruciate ligament (ACL) is injured. The human body has a multitude of protective structures that aid in the mitigation of injury. There are bones, tendons, ligaments, muscles, and fat or soft tissue that aid in the protection of a particular body part. Simply knowing the strength of the material is not always sufficient to understand the mechanics of injury. The actual stresses and forces must be directly applied to the particular material in order to produce the injury. In general terms, these forces and stresses have to exceed those that injure because of the protective structures of the surrounding biological materials. The mechanics of the movement that causes the injury must also be well understood. By performing calculations on a variety of injury scenarios, the investigator can gain great insight into the validation process for the various injury potentials.

Wayne State University and the Society of Automotive Engineers (SAE) as well as the National Highway Traffic Safety Administration (NHTSA) conduct many tests in order to assess the potential for injury. These ongoing tests have led to the development of seat belts and airbag systems. However, even with these systems in place, injuries do occur as a result of variations in humans and some improperly designed systems. A case in point is the original design for frontal crash airbags, which were very rapidly deployed and had a tendency to injure smaller individuals and children. Subsequent tests produced modifications to these systems and the implementation of different standard for children riding in automobiles.

Calculations performed on crashes that occur or in events that injure people are very good predictors of injury. Another set of calculations that have placed limits on head injuries is the Head Injury Criterion (HIC), which is part of the Federal Standards. Again, extensive testing brought about the lowering of the HIC standard from a value of 1,000 to a new lower level of 700. Continued testing brings about these changes in standards and helps to validate the potential for not only injury but also the human tolerance to injury.

Determine Human Tolerance Levels

Humans come in various sizes and shapes. Disregarding people who are outside of normal levels, those that are severely thin or overweight, a range for normal variability in humans is established. With respect to height, we would place this range between 4 ft 8 in. for small women and 6 ft 6 in. for tall men. Of course there are humans outside of this range, but they are exceptions and normally aberrations to the human condition. People shorter in stature normally suffer from a variety of maladies such as dwarfism. According to the medical literature, the definition of dwarfism is for people with a stature of less than 4 ft 10 in. Our definition is somewhat different to account for all people in the world, not just those from Western civilization. People taller than 7 ft in stature may be suffering from gigantism or acromegaly. Usually these two conditions are used interchangeably but they are not the same. Gigantism is caused by excessive growth during childhood or puberty, whereas acromegaly is a disease of adulthood. Both conditions generally result from tumors to the pituitary gland.

Weight is another factor that has to be considered. We place the upper limit on weight for both men and women to be around 260 lb except for white women that is in the neighborhood of about 230 lb. The lower limit for men is around 120 lb while that for women is about 105 lb. These rough values are age dependant. The charts that follow show the variability in the weight distribution for both white and black men and women according to age groups. Figure 5.1 represents white men's weight distribution and Figure 5.2 represents black men's weight distribution. Figures 5.3 and 5.4 represent the weight distributions for white and black women, respectively. The upper curves represent the weights at the 95th percentile and the lower weights at the 5th percentile. The 95th percentile indicates that 95% of the population weighs less than the given weight. The fifth percentile represents that only

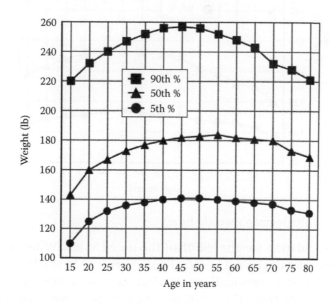

Figure 5.1 White men's weight by age.

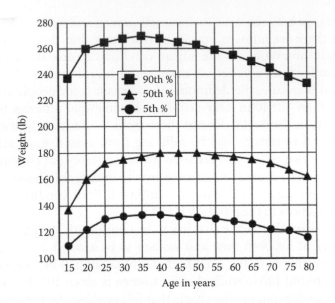

Figure 5.2 Black men's weight by age.

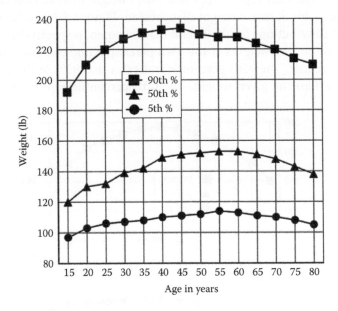

Figure 5.3 White women's weight by age.

5% of the population weighs less than the stated value. When a particular weight of an individual is not known, these charts can be used for parametric analysis to perform computations. A wider variability may be achieved by varying the weight from 100 to 300 lb, which would include roughly 99% of the human population. Although these graphs do not include all races, they

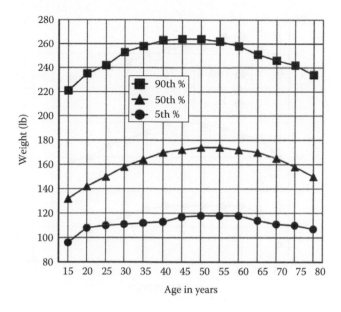

Figure 5.4 Black women's weight by age.

include the large percentage of the human population. Other races generally fall within the bounds represented by the curves shown.

Human height is another consideration that may be accounted for because of its relationship to weight and the dimension of a particular body part. It is important to point out that it is not feasible to measure, in most cases a particular dimension of a body part of an individual without imaging. Medical imaging may be available, but it is not readily accessible to the forensic engineer, which is especially true in the case of an individual making injury claims that cannot be substantiated. When an actual event injures an individual, there is generally no need to perform biomechanical calculations unless there is an attempt to place blame on, for example, a particular piece of equipment. The case of airbags comes to mind of this type of analysis. Figure 5.5 represents the 95th and the 5th percentiles for white men and black men's heights are in Figure 5.6. Figures 5.7 and 5.8 represent the weight distributions for white and black women. All of the graphs show the weight and height distributions according to age. As would be expected, the weight distributions for both men and women vary considerably as the people age. The curves have a greater bell-shaped peaking around the age of 45. In contrast, the height curves are much flatter between the ages of 20 and around the age of 40. There is a sharp increase of stature before the age of 20 when people are reaching mature height. After 40 all people tend to lose stature by a couple of inches to the age of 70. This age-related stature decrease is mainly due to vertebral disc compression resulting from the effects of gravity on age-weakened structures.

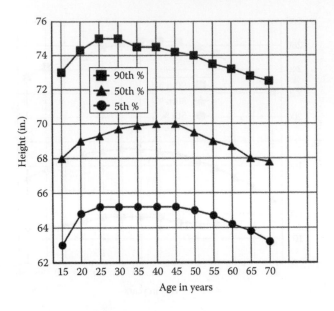

Figure 5.5 Black men's height by age.

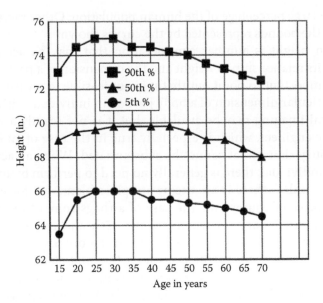

Figure 5.6 White men's height by age.

Most testing of human tolerance levels have been performed on relatively healthy and robust individuals. However, the literature is somewhat lacking on the effect of age, gender, and frailty of the individuals. By performing calculations on injury potential, a better picture should emerge on injury distribution by these parameters.

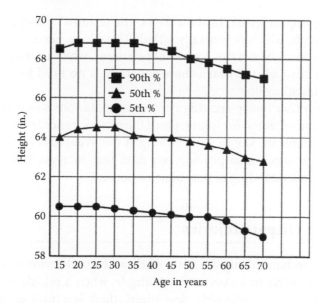

Figure 5.7 White women's height by age.

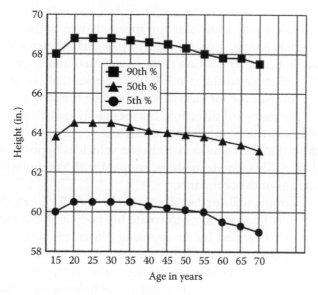

Figure 5.8 Black women's height by age.

Correlate Computations with Injury Potential

Computations should always be carried out to determine the forces and stresses that are produced when a human is injured. Of course, in some instances, computations do not need to be carried out. For example, if a sky-diver jumps from a plane and the parachute fails to open, there is no need to compute the impact speed. In these cases, the terminal velocity is approximately 120 mph, which is generally lethal. There have been, however, cases when the skydiver did not die. As another example, when a person is impaled by an object through the heart, there generally is no need to compute the force of the impaling object.

For most other injury-creating events, computations will either support or discredit the potential for injury. Once human tolerance levels are known, the computations are key in determining if the event was in consort with injury. Computations also provide raw data that increase knowledge and expand our understanding of how injuries occur.

Validate or Dispute Injuries

Forensic biomechanical engineers are often asked to determine if a particular injury is possible in an event. For example, when a vehicle is rear ended by another vehicle at a relatively slow speed, there is a limit as to the types of injuries that may occur. Sometimes occupants of the struck vehicle may claim that their knees struck the dashboard of the vehicle. A reconstruction of the collision may show that the speed change of the struck vehicle was approximately 5 mph. Physical laws, in particular Newton's third law of motion, state that for every action there is an equal and opposite reaction. Applying this principle to the two vehicles implies that the striking vehicle develops a force in the forward direction while the struck vehicle develops a force in the rearward direction. Further applying this principle to the occupants indicates that at impact, the occupants are thrust rearward against the seat back so that their motion within the vehicle is opposite the location of the dashboard. This simple analysis, coupled with computational values, dispute the claims of the occupants.

A sports example will further illustrate the need for analysis. During a football game, a runner attempts to rotate his body and change directions laterally when he is tackled by an opposing player. He suffers an anterior cruciate ligament injury to one of his knees. A lawsuit follows making claims against the opposing player and against the maker of the synthetic turf that covers the field. The two factors involved are as follows: (1) What is the slip resistance of the shoe interface with the synthetic turf and (2) was the tackle proper with respect to the forces that were generated?

One of the most common injuries that is claimed in vehicular collisions is "whiplash." This claim is generally created in rear-end collisions but may occur in t-bone collisions or frontal impacts. The scientific literature indicates that these injuries occur to the cervical portion of the spine when the neck whips from the collision causing bulging or rupture of the discs. The strength of these structures is well known so that forces and accelerations required to produce these injuries are correlated to speed changes in excess of 12 mph. Certainly, below speed changes of 10 mph, the scientific literature does not support whiplash-type injuries.

Other common injuries claimed in vehicular collisions include rotator cuff injuries, lumbar and thoracic spine injuries, carpal tunnel injuries, and jaw injuries, known as temporomandibular joint injuries (TMJ). It should be pointed out that rotator cuff injuries are associated mainly with sports and overhead movements of the arms, and occur mainly because the shoulder joint wears out as a result of usage and age. Lower spine injuries do not occur at lower speeds because the seat back and the seat belt and shoulder harness protect the occupant in motor vehicles. TMJ occurs as a result of the joint of the jaw wearing out and carpal tunnel afflicts people who overuse their fingers by typing or performing other repetitive tasks. These types of injuries are simply not associated with low-speed rear-end collisions between vehicles.

Design Safer Equipment

Over the years there have been a plethora of changes made to equipment that protects humans. These safety devices are known as personal protective equipment (PPE). The main components of the human body that are protected are the hands, the eyes, and the head. Fall protection is also a safety factor. Safety gloves of all forms reduce hand injuries by 27%. In electrical work, insulating gloves are a must to protect from shock or electrocution. Additionally, there are a variety of aprons, blankets, and protective equipment designed to isolate the worker from injury. One common safety device is the vest worn by police officers. These Kevlar vests worn by law enforcement and military personnel are designed to stop projectiles from penetrating the human torso, which is the most susceptible portion of the human anatomy to serious injury or death from penetrating projectiles. Many police agencies also equip their officers with helmets. Obviously, in order to ensure that the vests operate as required, the forces and penetrating power of the projectiles were needed to be calculated. Additionally, tests to ensure that the vests would keep projectiles from penetrating the material and lessen the impact of the force generated upon impact are needed to be performed. Brightly colored reflective vests are also required for conspicuity in any industry that exposes workers to traffic and moving machinery.

Every day approximately 200 workers sustain eye injuries. Eye protection is a must in any industry that produces particulate matter to become airborne. Welding, grinding, wood work, and similar industries require eye protection. Eyes should also be protected from the damaging rays of the sun.

In 2010, there were 2.5 million brain injuries ranging from mild concussions to severe injuries leading to death. These injuries were related to physical activity in sports such as football at the youth level to the professional level. They occur in baseball and soccer. If you are old enough, you remember that baseball players only relatively recently have been required to wear helmets while batting. Construction workers of all types are required to wear hard hats to protect themselves from falling objects around the work site. Recently, it has come to light that the helmets used in football may not adequately protect the players so that research efforts in this direction are taking place. For decades, both baseball catchers and umpires behind the plate have had significant impact protection from the ball because they are in line with the path of the ball. Let us see why a baseball can be very dangerous.

Baseballs have an average circumference between 9 and 9.25 in. They weigh between 5 and 5.25 oz. Let us say the balls weigh 5/16 of a pound or 0.3125 lb. Professional pitchers can throw the baseball in excess of 100 mph or 146.7 ft/s. The kinetic energy generated is given by

$$K_e = \frac{1}{2}\frac{w}{g}v^2 = \frac{1}{2}\frac{(0.3125)}{32.2}(146.7)^2 = 104.4\text{-ft-lb.} \tag{5.1}$$

To get greater insight, we can calculate the force that is produced by being struck with a baseball, which yields the pressure produced upon impact. When the ball strikes the batter, it decelerates from about 100 mph to zero in a short distance. Depending where the player is struck, that distance can vary considerably. On the head the thickness of the tissue is quite small, say 1/4 of an inch, which is 1/64 of a foot or 0.0156 ft. The deceleration is calculated from

$$a = \frac{v^2}{2d} = \frac{(146.7)^2}{2(0.0156)} = 689{,}772 \text{ ft/s}^2. \tag{5.2}$$

The force is the product of the mass and the acceleration or

$$F = ma = \frac{w}{g}a = \frac{0.3125}{32.2}(689{,}772) = 6{,}694 \text{ lb.} \tag{5.3}$$

The pressure exerted by the impact is the force divided by the area. The area of the ball that strikes the head would vary approximately from 1/4 to 1/2 of the circumference of the baseball. Let us just assume that the entire circumferential area makes contact. The area would then be about 6.8 in.[2] The pressure is then about 984 psi. For half the circumference impacting the head we get 1968 psi and for one-fourth of the area we get 3936 psi. These values clearly indicate the danger of injury that a baseball can inflict and display the need for head protection.

The second leading cause of unintentional death in the United States is produced by falls. Falls create a great danger as we saw in the simple calculations of Chapter 3. These calculations show the speed attained by a falling object depending on the distance that the object falls. The Code of Federal Regulations (CFR) requires that workers must have devices that prevent the workers from falling while working above a certain height. This is because falls above a certain elevation have the potential to produce serious injury or death. Fall protection is required under CFR 29, Part 1926.501 (b), which states that workers above an elevation of 6 ft need that protection. The protection may be by railing, nets, or personal protection harnesses. Recalling the speed attained from a fall distance of 6 ft to be 13.4 mph, we begin to see the reasoning used to protect humans from falls or collisions. These calculations are also evident in the deployment of airbag systems, which are designed to inflate when a speed change is produced in the 12–14 mph speed change.

Design Safer Machines

Mechanized equipment has a great potential to cause injury to humans. This equipment includes all types of machines used in industry, commerce, transportation, and construction. These machines may be stationary or they may travel. In general, the movement of mechanized equipment, which is generally constructed of hard and stiff components, is powered by engines or power sources that develop significant energy. We have already seen the power produced by one horsepower that is certainly very capable of causing great injury or death. Many of these machines and equipment produce scores, if not hundreds of horsepower. In the design of equipment, calculations need to be conducted so that the workers or users of the equipment are guarded and protected from injury. Over the past century, a variety of professional organizations and the Federal Government have instituted safe practices and procedures to address the safety issue. The capability of the machine to create injury must always be analyzed with the potential for injury.

Machine safety has two goals. One goal of a safe machine is that it will not injure a user or worker while operating the machine in a proper manner for which it was designed. The other goal is to recognize that a machine is potentially dangerous to operate so that safety features are built into the design to guard or protect the user. A very simple example of machine guarding is a handheld circular saw. The saw blade actually has a protective shield that retracts as the saw cut on the material is activated. When the cut is finalized and the trigger is deactivated, the inertia of the saw blade keeps the blade rotating. Consequently, the blade guard repositions itself to protect the user. The other classical example of proper use of a product is driving a vehicle on an Interstate highway. The speed limit is 70 mph, but the average vehicle is fully capable of traveling 100 mph. Traveling in excess of the posted speed limit is a danger to you and to other drivers. Road designers have made the appropriate computations to determine these safe speed limits for the average driver. Similarly, car designers have included safety devices on the machines such as seat belts and airbag systems to protect the occupants in the case of a collision. However, these safety devices are intended for the proper use of the machine, that is, traveling at a safe speed. All bets are off if you crash going 100 mph.

Biomechanical Terminology

6

Introduction

Most of us who have not been trained in biological sciences, medicine, and anatomy simply refer to various parts of our bodies as the head, chest, back, arms, legs, and various other parts with common terms that everyone understands. Fortunately, anatomical expressions are much more definitive and concise. To simply say that an injury occurred to the leg is not sufficient to accurately describe the injury and the processes that caused the trauma. In this context, we must explain the anatomically correct portion of the body that sustained the injury. The particular structure of the human body is described by an anatomical term. The human body has been quantized into several regions, which allow the use of proper terms to describe the location of the structure of interest. In this manner, proper assessment and computation can be made to mathematically and mechanically describe and quantify the injury or its propensity. Additionally, when reviewing a medical diagnosis from a health provider, the injury will be explained with a medically and anatomically succinct description.

As examples of the disparity and misuse of terms, relative to anatomical descriptions and the use by the general population, consider the term "leg." Anatomically, the term leg means the area of the lower limb between the knee and the ankle. It does not mean the upper portion, which is commonly referred to as the thigh. It also does not mean the entire leg between the hip and the ankle. More specifically and anatomically, the upper leg refers to the upper third of the tibia, which is the part of the tibia attached to the knee. Similarly, the lower leg refers to the lower third of the tibia, which attaches to the ankle. In a similar fashion, the term "arm" refers to the area between the shoulder and the elbow in correct anatomical terms. The area between the elbow and the wrist is correctly called the "forearm." The more explanatory location of a particular section is further described in this chapter as proximal, medial, and distal.

Figure 6.1 shows a frontal and a side view of a female describing the common terms used to identify various parts of the anatomy. It is evident from Figure 6.1 that a common description of the body parts is not adequate in identifying the portion of the body that has been injured and the mechanism that created the injury.

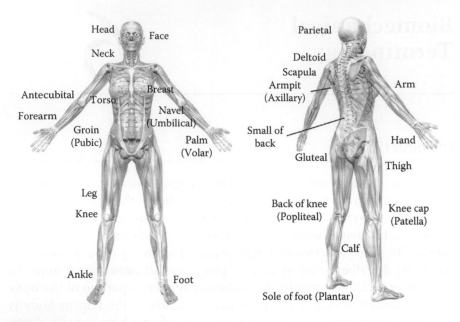

Head
Face
Neck
Parietal
Deltoid
Scapula
Armpit
(Axillary)
Arm
Antecubital
Torso
Breast
Forearm
Navel
(Umbilical)
Groin
(Pubic)
Palm
(Volar)
Small of
back
Hand
Gluteal
Thigh
Leg
Knee
Back of knee
(Popliteal)
Knee cap
(Patella)
Calf
Ankle
Foot
Sole of foot (Plantar)

Figure 6.1 Female anatomy.

The various body parts utilizing common terms are outlined in Figure 6.2 with some more descriptive terms that further identify the various body parts.

Technically trained individuals would automatically apply a coordinate system to the human body in terms of a geometrical description. The most common system would be in terms of Cartesian coordinates with directions x, y, and z. The center of mass of the body would be the origin at approximately the location of the navel inside of the torso. The positive x-direction would be to the front (ventral or anterior), and the negative x-direction would be to the rear (dorsal or posterior). The positive y-direction would be above the horizontal plane of the navel (up or superior), and the negative y-direction would be below the navel (down or inferior). The positive z-direction would be to the right (outside or lateral and inside or medial), and the negative z-direction would be to the left (outside or lateral and inside or medial). Although this geometrical description can be quite illuminating, it lacks the descriptive location of a particular body part in the human anatomy. Consequently, a set of anatomical coordinates much better describe the location of the body part in question. The anatomical description of the human body is represented by a set of three intersecting planes. These are the transverse, coronal, and sagittal planes. The sagittal plane that divides the body into the right and left sides is called the median or mid-sagittal plane. The plane that passes through the ears and divides the body from the front to the back is the coronal plane. Finally, the horizontal plane that

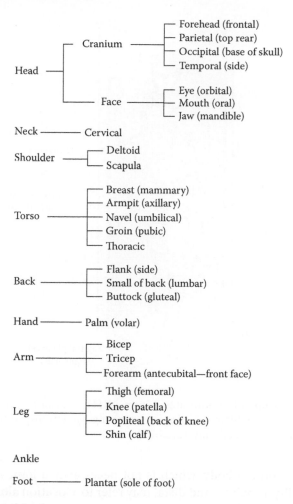

Figure 6.2 Body parts.

divides the body into the upper and lower portions through the navel is the transverse plane (see Figure 6.3).

Another method of locating a particular point on the body is to reference the close proximity of a known feature such as a knee, an elbow, a shoulder joint, or some other features. For example, the forearm can be said to be inferior to the elbow or that the elbow is superior to the forearm. With respect to the coronal plane, we can locate a body part as being toward the center (medially) or away from the center (laterally). For example, most of the ribs emanate from the breast bone (sternum) so a location that is closest to the sternum would be medial and further away would be lateral.

A further refinement with respect to location is internal or external, and proximal (near) or distal (far). The terms internal and external need no further explanation. However, the terms proximal and distal refer to a particular

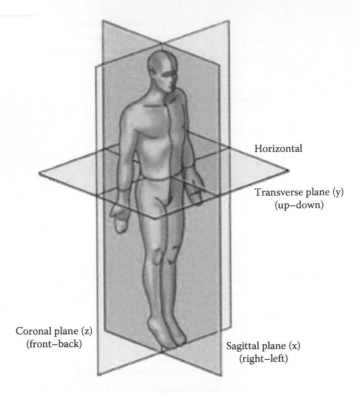

Figure 6.3 Anatomical coordinates. (From NIH's, National Cancer Institute, SEER Training Module, http://training.seer.cancer.gov/anatomy/body/terminology. html#directional, accessed May 12, 2015.)

reference point on the body, which is not necessarily a joint or fixed location. In actuality, proximal and distal may refer to a location along the length of the bone such as the humerus, which is accomplished by dividing the bone into thirds. The divisions of the humerus are P3, M3, and D3 denoting, respectively, the upper third (proximal, P) to the shoulder, middle third (middle, M) of the humerus, and the furthest third away from the shoulder (distal, D). If the section is the upper third of the humerus, it may also be described as the superior portion of the bone; or if it is closer to the elbow, it may be said to be inferior to the shoulder. The anatomical terminology is always described in terms of a standard position of the human anatomy. This position is always standing erect facing forward with the palms of the hand also facing forward. This is true regardless of the position of the body when inspected whether the person is alive or dead or is being treated. As a final description, the body may be described as being supine or prone. Supine means the body is on its back and prone means the body is on its chest.

Sometimes, it is convenient to describe the portion of the body with respect to the major components. These are the skeleton, muscles, tendons, ligaments, joints, internal organs, or the main external features such as the head, neck, torso, upper limbs, lower limbs, hands, and feet. The torso may be further divided into the thorax, which is the upper portion or the abdomen, which comprises the lower portion. Reference may also be made to a particular joint or point of attachment. We begin this refinement in position with the structural framework, which is the skeleton.

Skeletal Terminology

Our posture, shape, dimensions, strength, appearance, and functionality are dictated by the framework, which is comprised of our bones. There are over 200 bones of varying shapes and sizes in the human skeleton. Bones are generally classified in terms of their shape, development, region, and structure. In this section, we are not concerned to a great extent with the classification of bones in terms of their region or development. With respect to their development, these may be membranous bones, such as those of the face and skull, or cartilaginous bones, such as those at the end of the limbs, vertebral column, and thoracic cage. Another type of developmental bone is membrocartilaginous such as the clavicle, the mandible, and the occipital, temporal, and sphenoid bones. Classification of bones in terms of their region falls into two categories, axial and appendicular skeletal bones. Axial bones include the skull, vertebral column, and thoracic cage. Appendicular bones are generally those of the limbs.

Rather, the shapes and structures of bones are more essential for biomechanical calculations and analyses. Long bones are those whose length is greater than their width and that their shaft tends to widen at the ends. The ends are generally attached to other bones. These include the leg bones, arm bones, and finger and toe bones. There are short bones that have more or less a cube-like structure. This class of bones may be of any shape and are generally restricted to the bones of the hands and feet such as the carpals and tarsals. There are flat bones such as those of the skull and the ribs. These are also generally curved. There are irregular bones such as those of the face, vertebrae, scapula, and hip. Note that the scapula is also mostly flat. In fact, many bones cannot be completely ascribed to a particular shape classification. The fifth classification of bones is sesamoid bones. These are short and imbedded in a tendon such as the patella, pisiform (the smallest of the carpals), and two small bones at the base of the first metatarsal. Some flat irregular bones are sometimes classified as pneumatic bones. These bones are essentially hollow making them lightweight, such as the sphenoid, maxilla, and ethmoid, which comprise the major portion of the skull.

The structure classification of bones is sometimes critical in terms of biomechanics. Bones are all composed of the same material and only differ in their structure or arrangement pattern. Structurally, bone may be classified as macroscopic or microscopic. Macroscopic refers to the large-scale arrangement of the structural elements. There are two types of macroscopic classification of bones: compact and spongy bones. Compact bone has a greater ratio of bone to bone space. Spongy bone has more space than bone material. In terms of the microscopic classification, the bone may be lamellar or fibrous. Most mature human bone is lamellar. Fibrous bone is restricted to the fetus.

Many times it is necessary to compute the forces that damage cartilage. In this context, we need to further explain cartilage, its type and its function. Cartilage serves as a type of connective tissue that is found in various parts of the human anatomy in terms of its functionality. Cartilage is found at the ends of some bones but is not as hard as bone and is more flexible and elastic. It has no blood vessels, nerves, or lymphatics. Some cartilage is surrounded by the perichondrium membrane that aids in its regeneration. Articular cartilage has no perichondrium membrane so that its regeneration after injury is severely hampered. Cartilage requires a significant amount of time to heal and may not do so completely.

There are three types of cartilage: hyaline, elastic, and fibrous. Hyaline cartilage is found at the ends of long bones and is articular. Hyaline cartilage essentially acts as a lubricating and cushioning membrane between the articulating bones. Fibrous cartilage is much more fibrous than hyaline cartilage and is found between the sternum and the clavicle. Elastic cartilage is much more supple and is found in the eustachian tubes and the epiglottis.

There are several terms associated with bones that are descriptive of the feature of the bone. Condyle is a rounded projection for articulation at a joint, such as the condyles at the base of the femur that articulate with the tibia. The crest of a bone is a ridge to which muscles attach. Meatus refers to a tube-shaped opening of a bone, such as the external auditory meatus of the ear. Foramen refers to the hole in the bone. For example, the bones of the spine have a hole through which the spinal cord passes. Other holes are also present in the spine, which allow nerves and blood vessels to pass through to various parts of the body. Fossa refers to a shallow depression in a bone. Head is the rounded end that extends from the neck and fits into a joint as with the femur and hip joint. Sinus refers to an air-filled cavity in a bone as with the maxillary sinus of the upper jaw. The spinous process or spine is a sharp projection for the attachment of muscles to bones. Trochanter, of which there are two, is a projection at the proximal end of the femur where the muscles attach. These are the greater and lesser trochanters referring to their relative sizes. Tubercle is a smaller rounded projection for muscle attachment of the humerus. There is a greater and lesser tubercle at the proximal end of the humerus. Tuberosity is a large rough projection for muscle attachment as with the tuberosities of the ischium.

Joints

The junction between bones and cartilage comprises a joint that permits relative movement between the respective surfaces. Some joints allow greater movement than stability and vice versa. The shoulder joint is more mobile than the hip joint, but the hip is more stable than the shoulder. Some joints are immovable and are there to allow growth to proceed before and after birth. All joints have at least two surfaces: one male and one female. The male surface is larger and convex, and the female surface is generally smaller and concave. Simple joints have two articulating surfaces, and compound joints have more than two surfaces. Consequently, the joint may have uniaxial, biaxial, or multiaxial movement in all three directions including intermediate positioning. The joint may not also move. There are three types of joints relative to the degree of movement: fibrous, cartilaginous, and synovial. Fibrous joints permit essentially no movement between the bones as with the skull bones. Cartilaginous joints permit some very limited movement. The classic examples are the joints of the spine, which are referred to as the intervertebral disks. The third type of joint is the synovial joint with multiple degrees of freedom. One example of the uniaxial synovial joint is the elbow. Biaxial synovial joints may be either condyloid, ellipsoid, or saddle. The knee is the classic condyloid joint. The multiaxial joints or sphenoidal synovial joints are the shoulder and the hip.

The different varieties of movement are broken down into 10 basic movements: flexion, extension, abduction, adduction, pronation, supination, inversion, eversion, rotation, and circumduction. Flexion and extension refer to the bending and straightening of the joint such as that of the knee. Abduction and adduction refer to movement away and toward the midline of the body, respectively, as when the arm is raised or lowered laterally. Pronation and supination of the wrist refer to palm down and palm up, respectively. Inversion and eversion of the ankle means that the plantar portion of the foot is rotated toward the centerline or the body and away from the centerline, respectively. In rotation, the bone moves around a longitudinal axis, and in circumduction a long bone circumscribes a conical space. An example of rotation is the movement of the head by rotating the neck from side to side. An example of circumduction is the movement of the arm or wrist so as to create the conical surface. Many of the movements of the joints are complex and involve a combination of the different varieties. Figure 6.4 outlines these movements.

Some terms may have a prefix *hyper* to indicate excessive motion. The term may refer to excessive or rapid movement or both. One example is in the movement of the neck. If the neck is moved so that the chin contacts the chest, it is being hyperflexed. In the opposite direction, the neck is being hyperextended as

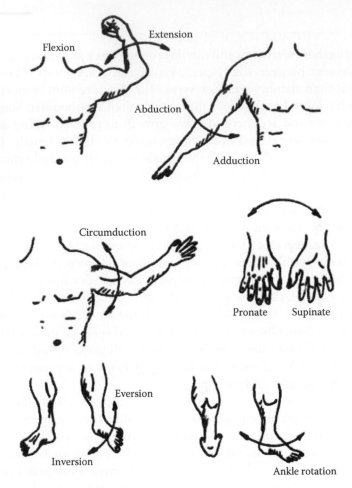

Figure 6.4 Body movements.

in the case of a rear-end collision leading to the common term *whiplash*. A common term associated with occupants of a vehicle that was involved in a rear-end impact by a treating medical expert is *cervical hyperextension*. Other common terms sometimes encountered are dorsiflexion, plantarflexion, opposition, and reposition. For example, dorsiflexion is the movement of the foot toward the shin, and plantarflexion is in the opposite direction. Opposition is when the thumb moves toward the other fingers, and reposition is the return movement.

Spine

A lot of emphasis is placed on the spine in terms of injury. The cervical spine has seven vertebrae. These are designated as C1 through C7 from the top-down. The thoracic spine has 12 vertebrae labeled T1 through T12.

The lumbar spine has five vertebrae designated as L1 through L5. The sacrum is composed of five bones that are fused together known as S1 through S5. The coccyx is comprised of four very small bones, which are the remnants of a tail. Each of the vertebrae is separated by intervertebral discs composed of flexible cartilage to allow movement and to act as shock absorbers.

Each disc has two main components. The outer portion is made of fibrous material that acts as a sack and contains the nucleus pulposus. Damaging the disc is referred to as prolapses, which causes some of the contents of the disc to leak out and put pressure on the spinal cord. Commonly, we see a term of *bulging disc* from the medical information of a medical provider. It is a well-known scientific fact that most people over the age of 40 have bulging discs. A diagnosis of a bulging disc for a person over the age of 40 does not necessarily mean that it was caused by a particular incident such as a vehicular collision. Discs tend to bulge as a result of age, usage, and gravity.

Muscles

Muscles are the mechanism of the human body by which force is exerted. Force is created when a muscle contracts, that is, it shortens. The muscles of the body are attached to tendons, which are then attached to the human skeleton. Muscles can only contract in one direction. As such, most muscles require an opposing muscle or group of muscles in order to counteract the movement that has been produced. A classic example is the bicep, which when contracted, moves the forearm toward the shoulder. In order to return the forearm to a straight position, the triceps need to be activated. Thus, most muscles are arranged in opposing pairs as with the two branches of the biceps muscles (biceps brachialis) and the three branches of the triceps muscles (triceps brachii). The general terminology for this dual action of the muscles is "primary mover" and "antagonist." Note that movement in one direction caused by a prime mover such as the bicep bundle in a curling motion may also be an antagonist in straightening the forearm.

Other muscles may be categorized as neutralizers. Neutralizers generally oppose the action of a prime mover and produce a motion that is different than that of a prime and antagonist pair. The motion produced by neutralizers is such that it would not be possible with the prime and antagonist pair alone. For example, biceps can flex the elbow and supinate the forearm. If only the elbow flexion is desired, the pronator teres, which pronates the forearm, would contract and counteract the supinator component of the biceps and only elbow flexion would occur. Muscles may also be categorized as fixators. These muscles stabilize a particular motion at a particular orientation. For example, the wrist may be rotated with the forearm up and down, at an angle, or rotated through the shoulder. This movement is achieved by a combination that requires primes, antagonists, fixators, and neutralizers. As such, any particular muscle may act as one of the four distinct categories.

Injury Terminology

Common terms, such as break, cut, and bruise, are generally not encountered when perusing through the medical literature of a particular individual who has been injured or is deceased. More formal terms related to injury of the human anatomy are used. We define the most common terms as follows:

    Abrasion—An injury caused by rubbing or scraping of the skin.

    Anoxia—An abnormally low amount of oxygen in body tissues causing mental and physical disturbances as a result of *Hypoxia*.

    Aphasia—An abnormal neurologic condition in which language function is defective or absent relating to an injury to the cerebral cortex of the brain.

    Asphyxiation—To cause death or to lose consciousness by impairing normal breathing by breathing a noxious agent or by suffocation.

    Avulsion—Tearing away or forcible separation of a bodily structure or part as a result of either injury or a surgical procedure.

    Basal skull fracture—Fracture or break at the base of the skull.

    Bursitis—Inflammation of a bursa, especially in the shoulder, elbow, or knee. Also called *bursal synovitis*.

    Coma—A state of profound unconsciousness in which a person cannot open eyes, obey commands, or speak words that can be understood.

    Compound fracture—A fracture in which the broken bone or its fragments are exposed through a *laceration* (wound) in the skin.

    Concussion—A mild brain injury often caused by a blow to the head or a sudden, violent motion that causes the brain to bump up against the skull. Also called *traumatic brain injury* or *closed head injury*.

    Contrecoup—Occurring to the opposite side as when an organ accelerates rapidly striking the opposing surface and ricochets back. Generally reserved for brain injuries but may be applied to other organs.

    Contusion—An injury from a blow with a blunt instrument in which the subsurface tissue is injured but the skin is not broken. A bruise.

    Dislocation—Displacement of a body part, as the temporary displacement of a bone from its normal position. *Luxation*.

    Dismemberment—To amputate a limb or a part of a limb.

    Edema—Swelling that happens when too much fluid collects in the body's tissues or organs.

Exsanguination—Death by bleeding.

Fracture—A break, rupture, or crack of a bone or cartilage. *Comminuted fracture* results in more than two fragments and is caused by a blow or a twisting force. *Incomplete fracture* is one that does not completely traverse the bone. *Stress fractures* occur from repetitive physical activity.

Hematoma—A localized swelling filled with blood resulting from a broken blood vessel. A circumscribed collection of blood, usually clotted, in a tissue or organ.

Hemiplegia—Paralysis affecting only one side of the body.

Hemorrhage—An escape of blood from a blood vessel in excessive amounts. Also called *hemorrhea.*

Hypertension—High blood pressure.

Hypotension—Low blood pressure.

Hypoxia—Inadequate oxygenation of the blood.

Impacted fracture—A bone fracture in which one of the bone fragments is driven into another fragment. Generally, these fractures consist of several pieces that have been broken.

Innominate—Having no name.

Laceration—A jagged wound or cut. The process or act of tearing tissue.

Luxation—Dislocation as in a joint.

Paralysis—The inability to move a group of muscles.

Prolapses—Damage to discs.

Puncture—To pierce with a pointed object such as a needle. A hole or depression made by a sharp object. Also called *centesis.*

Septic/sepsis—Infection due to germs in the patient's blood.

Sprain—An injury to a ligament when the joint is carried through a range of motion greater than its normal range without *dislocation* or *fracture.*

Strain—To pull, draw, or stretch causing injury. To injure or impair by overuse or overexertion. To wrench, twist, or other excessive pressure, causing injury.

Tendinitis—Inflammation of a tendon.

Tenosynovitis—Inflammation of a tendon and its enveloping sheath. Also called *tendinous synovitis* or *tendovaginitis.*

The majority of the ailments described earlier are not as severe as lacerations, luxations, or fractures. These inflammatory conditions on the muscles, tendons, and ligaments are most often resulting from overuse. Repetitive acts such as typing, throwing, lifting heavy objects, or work-related activities will lead to muscular fatigue. Once the muscles fatigue, their ability to counteract forces is diminished. At that point, the adjoining structures such as tendons, ligaments, and bones must try to carry the load resulting in inflammation syndromes.

A common sport-related injury that most people have experienced is a sprained ankle. In fact, one does not have to be involved in a sport activity to suffer from a sprained ankle. The sprained ankle occurs when the foot is inverted causing the foot to turn in causing excessive force to be placed on the lateral ligaments that hold the ankle bones together. These ligaments are not very robust and are thus prone to injury. Sprained ankles have three degrees of injury. The first degree simply means that the ligaments are over-stretched. The second degree involves approximately one half of the ligament tears. In a third degree ligament sprain, the tear is complete.

A more severe ankle injury involves a fracture. One type of fracture occurs when an ankle bone breaks. Another type of fracture involves an avulsion type fracture where the ligament pulls a portion of the bone from its attachment. This condition occurs when the ligament is stronger than the bone attachment. Fractures can only be diagnosed through x-rays or other diagnostic imaging techniques unless they are compound types. Recall that compound fractures allow the particular bone to protrude through the skin.

Basic Elements of Anatomy

7

In order to analyze injuries to the human body, we must first discuss the elements that are involved. The logical sequence for analysis follows a standard beginning with the framework, which is the skeleton, and then proceeding with tendons, ligaments, muscles, skin, and organs. The approach for all these elements is from the top-down, that is, from the head to the feet. Various portions of the human anatomy are not covered because they either are not involved with trauma as a primary injury or are not susceptible to biomechanical analysis. For example, if a puncture wound from a lance or a knife penetrates the abdomen causing a laceration of the intestines, there is no reason to calculate the forces necessary to perforate the intestine itself. It is sufficient to calculate the forces that penetrated the outer walls of the abdomen. On another note, traumatic brain injuries can be created without a penetration of the skull per se. In fact, it is important to quantify the forces that produce these closed head injuries (CHIs). It is simply recognized that penetrating injuries from firearms produce sufficient energy to enter, penetrate, and in many instances exit the human body.

Bones

There are over 200 bones in the human skeleton. We will not attempt to name all of them because many bone names are somewhat repetitive as with the bones of the hands and feet. Figure 7.1 shows the front and rear views of the human skeleton with the major bones highlighted. A list of the bones in various parts of the skeleton is shown in Figure 7.1.

Bones of the head include the ethmoid, frontal, occipital, parietal, sphenoid, and temporal. The facial bones are the hyoid, lacrimal, mandible, maxilla, nasal, palatine, turbinator, vomer, and zygomatic. For our purposes, we only include the ones listed as follows:

Frontal bone
Mandible
Maxilla
Occipital bone
Parietal bone
Temporal bone
Zygomatic bone

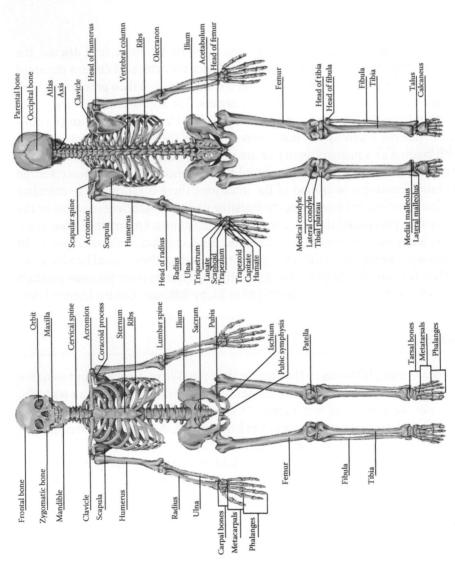

Figure 7.1 Human skeleton.

Bones of the Neck and the Chest

Acromion and coracoid, the projections of the clavicle

Atlas
Axis
Cervical vertebrae
Clavicle
False ribs
Floating ribs
Scapula
Sternum
Thoracic vertebrae

Bones of the Abdomen

Coccyx
Ilium
Ischium
Lumbar vertebrae
Sacrum

The arm, forearm, and hand—in the hand there are 27 distinct bones that allow for the complex dexterity of humans. This complex system allows for rotation, pronation, and supination of the hand as well as complex movement of the hand. The eight carpals are composed of two rows of four each attached to the radius and ulna proximally. The digits of the hand each contain three phalanges except for the thumb that contains two.

Carpals
Epicondyle
Epitrochlear
Humerus
Metacarpals
Olecranon
Phalanges
Radius
Ulna

The bones of the leg, ankle, and foot—in the foot there are also a variety of bones similar to the hand. These include the metatarsals and phalanges as well as the bones of the mid foot.

Calcaneus
Femur

Fibula
Metatarsals
Patella
Phalanges
Talus
Tarsals
Tibia

Probably, the most critical injuries to the human anatomy deal with trauma to the head because humans can survive with the loss of legs, arms, hands, feet, and certain internal organs but not with severe trauma to the head and in particular the brain. The outer covering of the cranium is the scalp, which consists of five layers and is approximately 0.625 cm thick. These layers are the skin, the subcutaneous layer, the galea, the loose connective tissue, and the pericranium. Lacerations to the subcutaneous layer produce extensive bleeding as this layer is extensively supplied with blood. Lacerations extending to the pericranium can produce hemorrhaging, which is referred to as subgaleal hematoma. Injuries to the skin include abrasions, contusions, and lacerations.

Injuries extending beyond the scalp will affect the bones of the cranium. Injuries extending beyond the skin layers of the face will similarly affect the underlying bony structures. With respect to the cavity of the cranium, the brain is surrounded by three layers called the meninges, which are the dura mater, the arachnoid, and the pia mater. It is important to note that brain injuries need not involve a fracture or a depression of the skull. Such CHIs are also very significant and may involve death or at least severe injury. In the case of significant brain deceleration or rotation, the forces involved can cause death by bleeding from cranial surfaces or internally. Death by bleeding is referred to as exsanguination.

Table 7.1 represents the force required to fracture the bones associated with the skull according to SAE J885. These values can be correlated to the strength of bones in Chapter 8.

The table gives the range of values derived from testing and includes the mean and the range.

Table 7.1 Skull Fracture Force

Skull Bones and Fracture Force in Pounds		
Bone	Range	Mean
Frontal	470–2650	1000–1710
Mandible	184–925	431–697
Maxilla	140–445	258
Occipital	1150–2150	1440
Temporoparietal	140–3360	702–1910
Zygoma	138–780	283–516

When a fracture to the skull occurs, the bones may be exposed to the elements via a laceration in the skin which could lead to infection. In these compound fractures, if the meninges become infected, the clinical diagnosis is that of meningitis. If the fracture results from an external blow, the bones may be pushed inward so that the condition is referred to as a depressed fracture. Depressed fractures may or may not be significant from a clinical perspective. Injuries to the brain may occur from depressed fractures in which the portion of the fracture injures the brain causing hemorrhage and or a hematoma.

In a similar manner, fractures to the facial bones may or may not cause brain injury depending on the severity and the proximity to the brain. One exception to brain injury from a facial fracture may result from traffic crashes in which CHIs occur. CHIs are further discussed in "Head Injury Criterion" section.

Head Injury Criterion

Approximately 50% of all head injuries are caused by motor vehicle collisions. Falls account for 20%, physical violence for about 13%, and the remainder for sporting events and miscellaneous causes. It should be recognized that awareness has increased on the dangers of concussions in contact sports such as football at all levels. The severity of the injuries results in fractures, contusions, concussions, hematomas, and diffuse axonal injury (DAI). Brain or head injuries involve at least two and sometimes three separate and distinct collisions for which the forces need to be calculated. In a vehicular incident, the first collision occurs between the vehicles or the vehicle and a stationary object. The second collision occurs and involves the occupants with various structures of the interior of the vehicle. The third collision involves the collision of the intracranial structures with the bony interior of the skull. In cases where falls occur or when an object such as a pipe strikes the head, only two collisions occur.

Over the past half a century, data have been assembled concerning concussions and other severe CHIs. Scientists at Wayne State University and others (Lissner et al.) have developed a curve relating head acceleration versus time as represented in Figure 7.2.

It has been noted that the time of application of an acceleration to the head as well as the peak acceleration plays a significant role in CHIs. The curve in Figure 7.2 represents the head injury criterion (HIC). Versace in 1971 approximated the curve of the HIC by the following equation:

$$HIC = (t_2 - t_1)\left[\frac{1}{(t_2 - t_1)}\int_{t_1}^{t_2} a(t)d(t)\right]^{2.5} \qquad (7.1)$$

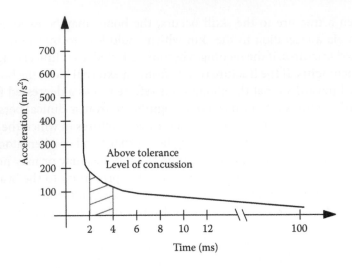

Figure 7.2 Closed head injury.

where $a(t)$ is the acceleration at each increment of time in g units (32.2 ft/s^2). The level of concussion occurs when the HIC number equals or exceeds 700. The prior level for the HIC was 1000 and was revised to the lower value as a result of further testing. Evaluating the integral, we obtain

$$\left(\frac{1}{t_2 - t_1}\right) \int_{t_1}^{t_2} a(t)d(t) = \frac{a}{2}(t_2 - t_1) \tag{7.2}$$

where a is now the maximum value. We may express the ratio of two HIC values given that

$$a = \frac{\nabla v}{(t_2 - t_1)} \tag{7.3}$$

as follows:

$$\frac{HIC_1}{HIC_2} = \left[\frac{\Delta V_1}{\Delta V_2}\right]^{2.5} \tag{7.4}$$

Nahum and Smith (1977) in their experiments on cadavers obtained pressure versus acceleration data of the brain for frontal impacts to the skull. These linear data are reproduced in Figure 7.3.

Ward et al. (1980) have shown that the threshold of intercranial pressure for brain injury is 24 lb/in.² or 1240 mm Hg. Above 34 lb/in.² or 1758 mm Hg, severe injury results. From Figure 7.3, we obtain a slightly different number

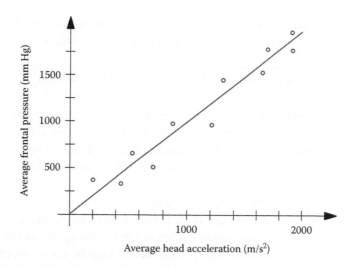

Figure 7.3 Frontal skull impacts.

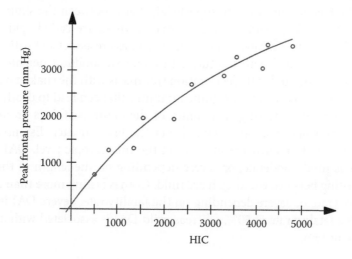

Figure 7.4 Pressure versus head injury criterion.

for the threshold of injury, which is 1293 m/s². The condition for serious head injury occurs at an acceleration of 1758 m/s² and above. As a result of these experiments, the original HIC value was determined to be 1000. These experiments also plotted the pressure versus the HIC value as shown in Figure 7.4.

If we consider the injury level at an HIC value of 1000 corresponding to 1750 mm Hg or 34 lb/in.², we can make a rough estimate of the impact of the frontal portion of the skull against a rigid surface that would result in serious injury to the brain. The associated acceleration is 1293 m/s². Stalnaker et al. (1977) have used high-speed cineradiography to x-ray the motion of the brain in frontal impacts to cadaver skulls. From these studies, they showed that the brain moves approximately 3 cm forward. We can then calculate the associated velocity required to produce serious head injury as follows:

$$v = \sqrt{2ax} = \sqrt{(2)(1293)(0.03)} = 8.9\,\text{m/s} = 19.8\,\text{mph} \qquad (7.5)$$

Concussions are broadly classified as injuries in which there is some loss of consciousness following some head trauma. These injuries do not present themselves as localized but rather as diffuse. Many CHIs can be classified as concussions. Generally, concussions immediately follow the loss of mental function rather than occurring at a later time. Concussions vary in degree from mild to severe. Contusions are associated with bruising of the brain's outer layer. If the contusion occurs at the primary point of contact, it is referred to as a coup injury. If the contusion occurs at the opposing side of the impact, it is known as a countercoup injury. Contusions may be coup, countercoup, they may be associated with a fracture, a herniation, or a combination of these. Hematomas involve injury to blood vessels causing hemorrhage leading to a hematoma. Hematomas may be superficial, that is between the skull and the dura mater and deep within the dura mater. These are called epidural and intradural, respectively. Most epidural hematomas are associated with impacts to the head associated with fractures. In contrast, subdural hematomas are more closely associated with deceleration trauma as with the shaken baby syndrome. Hematomas are a very significant injury that can lead to death.

DAI refers to a shearing-type injury of the brain itself. It is produced by excessive brain deformation, which shears the brain matter. Essentially, the brain is lacerated in a gross manner or at the microscopic level. DAI may be classified as mild, moderate, or severe depending on the length of the coma. Comas lasting between 6 and 24 h are mild. Comas lasting more than 24 h are either moderate or severe depending on the fatality rate. Severe DAI is associated with a fatality rate of 50%, whereas mild DAI is associated with a fatality rate of about 15%.

Spine

As we descend down the human anatomy, the next logical bony structure is that of the spinal column, which begins at the base of the skull and extends to the coccyx. The spinal column consists of a series of bones with an opening at the center, which allows for the extension of the brain stem which is the spinal cord. A simplistic but accurate explanation of the spinal cord is one that allows messages to be transferred to and from various portions of the human anatomy and the brain. As the spinal column descends, the spinal cord branches off at the cervical, thoracic, lumbar, sacral, and coccygeal regions. The bony vertebrae are cushioned by the vertebral discs and held

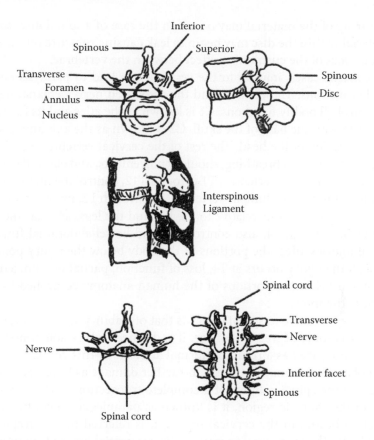

Figure 7.5 Vertebral segments.

together by the spinal ligaments, which are anterior and posterior longitudinal ligaments and the ligamentum flavum along with the interspinous ligament. Figure 7.5 shows various views of spine segments and parts.

Shown in Figure 7.5 are the spinous process, the inferior and superior facets, the transverse facet, the disc, the foramen through which the spinal cord passes, and the nerves emanating from the vertebral column. Also shown are the anterior longitudinal ligament and the interspinous ligament. The posterior longitudinal ligament and the ligamentum flavum are also shaded.

Spinal injuries can be broadly categorized as sprains, vertebral disc bulges or ruptures, or vertebral fractures. Some of these injuries can put pressure on the spinal column or on the nerves emanating from a particular vertebra. Sprains tend to affect the facet joints and ligaments, which can affect the discs. In a sprain, the soft tissue structures of the ligaments or the discs are stretched beyond the elastic limit of the material. When this happens, the material cannot return to its original shape and dimensions. Consequently, some tearing of the material may occur. In the case of a spinal disc, some of the material within the disc may begin to leak causing pressure on the spinal cord and a loss of the cushioning effect between the vertebrae.

In extreme cases, injuries to the vertebra can lead to paralysis and loss of bodily functions. The vertebra and the associated nerves emanating from them control all bodily functions. C1 is known as the atlas and its function is to articulate with the base of the skull. C2 is known as the axis and is associated with rotation of the head. The rest of the cervical vertebrae, C3 through C7, are associated with breathing, shoulder movements, and elbow flexion and extension. The thoracic vertebrae, T1 through T12, control the intercostal and abdominal muscles. The lumbar vertebrae, L1 through L5, involve a variety of movements and functions relative to the hip and the legs, whereas the sacral vertebrae, S1 through S4, also control foot and bowel/abdominal functions. Vertebral injuries affect the portions of the body below the injury point. For example, if the injury occurs at T4, loss of function, partial or complete, may occur at and below the portions of the human anatomy controlled by those portions of the spine.

A concept relative to spinal injury is that of stability. If a vertebral structure sustains some degree of damage, it is deemed to be stable if no further damage is produced as a result of normal activity. In this context, a broken section of a vertebra may not result in further damage as long as it does not impinge on the spinal column. If a complete transection of the spinal cord occurs in the thoracic region, it is known as paraplegia. In contrast, if the transection occurs in the cervical region, it is referred to as tetraplegia or quadriplegia. The term quadriparesis indicates partial loss of function in all four extremities. These partial cord injuries are referred to as incomplete spinal cord syndromes and account for approximately 90% of all spinal cord injuries.

The structure of the spinal cord is comprised of various tracts or cables if you wish. There are descending and ascending paths. There are also anterior, posterior, central, and lateral tracts and some are laminar in construction. An unstable spinal cord injury is one in which there is a real possibility that further cord damage will result from normal or perceived activity. Spinal cord injuries are generally produced by excessive loading in a variety of manners. These may be due to extension, flexion, rotation, torsion, tension, compression, shear, or a combination of these loading mechanisms. All of these mechanisms are addressed in the models that we have developed for spinal cord injuries.

More severe injuries to the spine are burst fractures and wedge fractures. Burst fractures are associated with compressive forces that are applied vertically down. Wedge fractures are also produced by compressive forces and generally apply to the anterior surfaces of the vertebra. Sometimes, the forces are horizontal in nature and may produce a dislocation or partial dislocation of the vertebra. Significant dislocations may result in fractures. Generally, the dislocations are partial rather than complete and are referred to as subluxation. Subluxations and fractures are sometimes related to CHIs as previously explained. For a more complete treatment of these injuries, refer to Baxt, Harris, Levy, Mahoney, Nahum, Clement, and others in the bibliography.

Some injuries to the neck area occur to the front where the epiglottis, thyroid cartilage (Adam's apple), and cricoid cartilage reside. These structures are important with respect to injury and death because of the abundance of blood and air vessels. Restriction of air to the lungs and blood to the brain results in significant injury and loss of life. Both the thyroid and the cricoid fracture at forces between 90 and 100 lb. These forces are easily obtained when a person is strangulated.

Muscles

The number of muscles of the human body depends on how they are grouped. Various sources will group pairs of muscles in a variety of manners depending on whether they are single or pairs of muscles. The range of the number of muscles is between 640 and 850, depending on the grouping. We are not concerned with many of the minor muscles of the body since we are dealing with major injuries to the human anatomy. Our grouping of the muscles is confined to the major muscles of the head and neck, the torso, the upper limbs, and the lower limbs. There are many muscles that control the scalp, ears, eyes, tongue, and soft palate. With respect to the head and neck, we are only concerned with the muscles of mastication, the pharynx, the larynx, the cervical, suprahyoid, and infrahyoid muscles, and the vertebra. These muscles are tabulated in the following, and the major muscles of the other parts of the body are also included. We have excluded some minor muscles or grouped them according to category. We have not listed any muscles of the hand or foot.

Muscle	Origin	Action
Masseter	Zygomatic arch and maxilla	Elevation and retraction of mandible
Temporalis	Parietal bone	Elevation and retraction of mandible
Pharyngeal	Cricoid cartilage and hyoid bone	Swallowing
Sternocleidomastoid	Clavicle	Flexion and rotation of neck
Longus and rectus capitis	Cervical vertebrae	Flexion and rotation of neck
Scalene	Cervical vertebrae	Elevation and rotation of neck and first rib

(*Continued*)

Muscle	Origin	Action
Rectus, semispinalis, longissimus, and splenius capitis	Cervical vertebrae	Extension, rotation, and lateral flexion of neck and head
Erector spinae	Thoracic vertebrae	Extension of vertebral column
Latissimus dorsi	Thoracic vertebrae	Pulling of forelimbs
Spinalis	Transverse/spinous process	Stability, extension, flexion, and rotation of head and trunk
Intercostal	Ribs	Inhalation, elevation, and depression of ribs
Serratus	Ribs and transverse processes	Depression and elevation of ribs and respiration
Abdominal	Ribs and pubis	Thoracic and pelvic stability
Oblique	Costae and iliac crest	Torso rotation and compression
Levator	Ischial spine and pubis	Support, urination, and defecation
Trapezius	Occipital protuberance	Retraction and elevation of scapula
Rhomboids	Spinous process	Scapula movement
Pectoralis	Clavicle and sternum	Clavicle, scapula stability, and movement
Deltoid	Clavicle and acromion	Shoulder flexion, extension, and abduction
Teres major	Scapula	Humerus rotation
Supra, infraspinatus, teres minor, and subscapularis	Scapula	Rotation, abduction, and stability of humerus
Biceps brachii	Scapula	Flexes elbow and supinates forearm
Brachialis	Humerus	Elbow flexion
Coracobrachialis	Scapula	Shoulder joint flexion
Triceps brachii	Scapula and humerus	Forearm extension and shoulder adduction
Anconeus	Humerus	Stability and extension of forearm
Pronator teres, flexor carpi, radialis palmaris longus, flexor carpi, and flexor digitorum	Humerus	Flexion, pronation, and abduction of forearm, wrist, and fingers
Pronator quadrus, flexor digitorum, and flexor pollicis	Radius and ulna	Pronation and flexion of forearm, hand, and thumb
Extensors	Humerus, ulna, and radius	Hand, finger, and wrist
Iliopsoas, psoas, and iliacus	Lumbar spine	Flexion and rotation of hip, thigh, and trunk
Gluteal	Ilium, sacrum, and ischium	Thigh and hip rotation, support, and abduction
Quadriceps and hamstrings	Femur, iliac, ischium, and greater trochanter	Flexion, rotation, and extension of knee

(Continued)

Muscle	Origin	Action
Adductors	Pubis and ischium	Adduction, flexion, and rotation of hip
Tibial and fibular extensors	Tibia and fibula	Extends, dorsiflexes, and inverts foot, toes, and ankles
Gastrocnemius, soleus, plantaris, flexors, tibialis, longus, and brevis	Femur, tibia, and fibula	Plantar flexion of knee, foot, and toes

Torso

We consider the torso as the portion of the body from the shoulders down to the legs. This segment of the body can be further divided into the upper torso and the lower torso. The dividing line is considered to be roughly the diaphragm, which separates the bony portion including the ribs, or thorax, from the lower portion that includes the abdomen. This line of distinction is somewhat arbitrary because we often think of the upper torso as the chest, which is protected by the ribs. In fact, the lower ribs also protect a portion of the organs of the abdomen that are below the diaphragm. The diaphragm is a muscular structure that aids in breathing and separates the lungs and heart from the lower abdominal cavity. The lungs are contained in the right and left pleural cavities. The cavity that contains the heart and its associated great vessels along with the esophagus is called the mediastinum. These organs are protected by the rib cage. Exterior to the rib cage are the breasts in men and women. The basic morphological structure of the breast is the same for both male and female children. At puberty, estrogen production in females produces breast development.

The rib cage consists of the sternum in the front and a portion of the spinal column in the rear. Normally, there are 12 ribs on each side. The top seven ribs connect to the sternum in the front and the associated top seven thoracic vertebrae. The lower five ribs are connected in the rear to the next lower five thoracic vertebrae. The top three of these lower five ribs are connected to the other ribs above via cartilage. The lower two ribs are floating in the front and are attached to the muscles of the abdomen. The top portion of the sternum is the manubrium, the center portion the gladiolus, and the bottom portion the xiphoid.

The lower portion of the torso, the abdominal cavity, contains the liver, pancreas, appendix, kidneys, spleen, bladder, gall bladder, ureter, renal glands, stomach, small intestine, and colon. The colon is further subdivided into the ascending, transverse, descending, and sigmoid sections. The sigmoid colon leads to the anus. Additionally, the reproductive organs of females and males are contained within the abdominal cavity except for the penis and testes of the male. These are located in the pelvic region of the lower abdomen. The following table lists the major organs and their function. Some relevant dimensions are also included.

Organ	Function/Facts
Brain	Center of the nervous system, controls all activities of the body (volume 970–1400 cm³ and weight 1.5–1.6 kg)
Heart	Muscular pump for blood providing oxygen, nutrients, and removing waste (12 cm long, 8 cm wide, 6 cm thick, and weight 250–350 g)
Lungs	Transport oxygen from the atmosphere to the blood stream and CO_2 out (length 25–35 cm, width 10–15 cm, and weight 0.8–1.2 kg)
Esophagus	Fibromuscular tube from pharynx to stomach (18–25 cm long)
Liver	Detoxification, protein synthesis, biochemical production, and digestion (1.4–1.7 kg in weight)
Pancreas	Endocrine gland producing insulin and other hormones for digestion (15 cm long)
Appendix	Pouch at small intestine colon junction. True function unknown (11 cm long and 8 mm in diameter)
Kidneys	Filters of blood, remove waste urine, maintain water, glucose, and acid balance (10–13 cm long and 125–170 g in weight)
Spleen	Blood filter and immune system functions (7–14 cm long and 150–200 g in weight)
Bladder	Collects urine from kidneys (5–13 cm long and 500–1000 mL in volume)
Ureters	Tubes from kidneys to bladder (3–4 mm diameter and 25–30 cm long)
Adrenal glands	Produce epinephrine, norepinephrine, and hormones (30 mm wide, 50 mm long, 10 mm thick, and 5 g in weight)
Small intestine	Absorption and digestion of food and minerals (4.6–9.8 m in length and 2.5–3.0 cm in diameter)
Colon	Absorbs water and rids of waste material (1.5 m in length)
Gall bladder	Stores bile for digestion and released into small intestine (8 cm in length, 4 cm in diameter, and 100 mL in volume)
Stomach	Between esophagus and small intestine that secretes enzymes and acids to aid in digestion (relaxed volume 45–75 mL and expanded 1000 mL)
Breast	Contains the mammary glands (weight from 500 to 1000 g)
Reproductive organs	Females (uterus 7.5 cm long, 5 cm wide, and 2.5 cm thick) and males (testes 2.5 cm wide and 5 cm long)

The varied nature of the torso allows for significant differences in the manner that it may become injured. The rib cage allows for the protection of the most vital organs from accidental injury for the most part. In order to injure the heart and the lungs, it is generally necessary to breach through the ribs. Recall that the rib cage acts as a protective shell, which flexes to a certain extent under loading. The cartilage attachments of the ribs to the sternum and the vertebra allow for this flexion along with the relatively flat design of the individual ribs. In terms of a beam structure, the individual ribs are superbly designed for strength and flexibility. Of course, evolution did not design the rib cage to be subjected to the forces produced by man in the form of weapons and devices such as automobiles.

There are two types of injury that greatly affect the upper torso. These are blunt impact and penetrating impact. Penetrating impact to the chest cavity may result from gunshot, puncture from objects such as knives, or extreme forces that fracture the rib structure. Blunt impact may be produced by a variety of objects such as steering wheels, hammers, or impacts from falls or large devices or structures. Blunt traumas, although they may not penetrate the rib structure, can be quite severe or lethal. Significant impacts to the rib structure can result in a fractured rib that can cause additional trauma such as a punctured or collapsed lung. Even if the ribs are not fractured, a secondary impact to the internal organs can cause death resulting from bleeding of the organ or its associated vessels. Internal secondary impacts generally result in hematomas, contusions, or tears. The aorta is particularly subject to tears resulting from blunt force trauma to the thoracic cavity.

The lower torso or abdominal cavity is much more susceptible to injury because of the lack of a rigid framework that affords protection. Somewhat similar in design to the thoracic organs, the abdominal organs are surrounded by the peritoneum. The peritoneum divides the abdomen into a frontal and a rear compartment. This partition includes the intrathoracic abdomen, which contains the liver, stomach, spleen, pancreas, and duodenum. The kidneys are also normally included in this region. The pelvic abdomen contains the digestive system and reproductive organs. Puncture or tear injuries to the abdominal cavity organs are subject to leakage of blood or of digestive contents increasing the possibility of sepsis. Blunt trauma to these organs also causes bleeding in particular to organs such as the liver and the kidneys. In contrast, hollow organs such as the colon or bladder are less subject to blunt trauma. The most common mechanisms of injury to abdominal organs include organ impact to bony structures, shear injury at attachment points, and burst injury.

Pelvis

The pelvis is composed of several bones, which are fused by adulthood. These bones form a flat structure that has a somewhat bowl appearance, which is concave anteriorly. The pelvis is symmetrical about the sagittal plane. The three main parts of the pelvis are the right and left bony structures and the middle portion composed of the sacrum and coccyx. The right and left protuberances are called the innominate bones. The upper part is the ilium, the midsection the ischium, and the lower portion the pubis. Please refer to Figure 4.12. The acetabulum is a semihemispherical structure on either side of the pelvis that accepts the head of the femur and is located inferolaterally.

Fractures of the pelvis are quite serious because they may affect the organs that are located within that region of the anatomy. Not only the organs but also a variety of blood vessels may be affected. Depending on the exact location of the fracture, the weight bearing ability of the pelvis may also be compromised. A luxation of the hip joint is a serious injury that can result in frontal clashes in automobiles. The luxation may be associated with damage to the acetabulum. In extreme cases, especially with older individuals, a hip replacement may be necessary. Fractures of the posterior of the pelvis are generally more severe because of the weight-bearing function of that portion of the pelvis. Anterior pelvis fractures have a greater propensity to cause associated damage to the internal organs.

Tendons and Ligaments

As previously described, tendons are the attachment of muscles to bones, whereas ligaments are the attachment of bones to other bones. Another type of connective tissue is fasciae, which connects muscles to other muscles. With respect to injury as a result of trauma, we are not concerned with injuries to fasciae because these injuries are generally concomitant with injuries to bones, muscles, ligaments, and tendons. A significant number of joint injuries include ligament and tendon damage as well as fractures.

Tendons are comprised of tough fibrous connective tissue parallel in construction primarily made of collagen fibers. Tendon names are associated with the particular muscle or muscle group. Tendon length varies for all the muscle groups from person to person. Some people may have short bicep tendons and long calf tendons. In general, muscle size is determined by the length of the tendon. The shorter the tendon, the larger the attached muscle. Tendons are not only a connective mechanism between the muscle and the bone but also have energy storing and spring-like characteristics. As with all biological materials, tendons exhibit stress–strain characteristics as is demonstrated in Chapter 8 and other sections of the book. As a rough estimate, tendons that have energy storage capabilities fail around the 12%–15% of the strain corresponding to a stress between 100 and 150 MPa. Recall that MPa is megapascal (mega is 1,000,000 and a pascal is a newton per square meter). Tendons that are designed to position the bones have lower strains in the range of 6%–8%. Studies have shown a direct correlation between lack of use of a tendon and muscular atrophy.

There is a misconception in the general population that athletes with exaggerated muscles such as body builders are much stronger than individuals with less musculature. This is actually not true because strength is mainly related to the attachments and not the muscle size. Tendons may be injured in a variety of ways. Paratenonitis is an inflammation of the sheet between the tendon and its sheath. Tendinosis is an injury at the cellular level, which is noninflammatory and may lead to rupture of the tendon. Paratenonitis with tendinosis is another form of injury to the tendon. Tendinitis is degeneration at the cellular level with inflammation and vascular disruption. These conditions may be due to a variety of factors such as age, weight, nutrition, the environment, training, and excessive forces from sports or loading conditions.

Ligaments are similar to tendons since they are both made of connective tissue. Ligaments differ somewhat from tendons in that they normally do not regenerate. Normally, the term ligament refers to articular ligaments in that they aid in the joining, movement, and stability of the bones. In a broader sense, ligaments may be classified as capsular, extracapsular, intracapsular, and cruciate. Capsular ligaments are a part of the capsule that surrounds synovial joints and act as reinforcement of the joint. Extracapsular ligaments aid in joint stability. Intracapsular ligaments also provide stability but are more viscoelastic and have greater range of motion. Cruciate ligaments, such as those in the knee, come in pairs of three.

The viscoelasticity of tendons indicates that they shrink under tension and return to their original length and shape when tension is removed. When a ligament is broken, it normally requires surgery to repair it. If not broken, in the case of a dislocated joint, the ligament will lose its elastic properties if not treated in a timely manner. Consequently, it is imperative to reset a dislocated joint soon after the event in order for the ligament to retain its elastic properties. Otherwise, the joint will be subject to future dislocations.

As with the sections on the bones and muscles, we do not include most ligaments of the foot and hand. The following is a list of the major ligaments of the human anatomy associated with the major joints.

Ligament	Function/Connection
Shoulder joint	
Acromioclavicular	Acromion to clavicle
Coracoclavicular	Coracoid process to scapula and clavicle
Glenohumeral	Humerus to glenoid cavity
Coracohumeral	Coracoid process to scapula and humerus
Transverse humeral	Greater to lesser tuberosity of humerus
Elbow joint	
Annular	Trochlear notch to ulna
Radial collateral	Humerus and radius
Ulnar collateral	Humerus to coracoid process and olecranon process
Wrist joint	
Pisometacarpal	Pisiform to ulna
Palmar ulnocarpal	Ulna to carpus
Palmar radiocarpal	Radius to carpal
Dorsal ulnocarpal	Ulna to carpus
Spine	
Interspinous	Connects from one spinous process to the next
Supraspinous	Connects entire vertebral column
Anterior and posterior longitudinal	Holds the vertebral bodies together
Ligamentum flavum	Connects the lamina of the spine
Hip joint	
Iliofemoral	Pelvis to femur
Pubofemoral	Pelvis to femur
Ischiofemoral	Ischium to trochanters of femur
Knee joint	
Anterior cruciate	Femur to tibia anteriorly
Posterior cruciate	Femur to tibia posteriorly
Lateral collateral	Femur to fibula laterally
Medial collateral	Femur to tibia medially
Ankle joint	
Anterior talofibular	Fibula to talus anteriorly
Calcaneofibular	Fibula to calcaneus laterally
Posterior talofibular	Fibula to talus posteriorly
Tibiofibular	Tibia to fibula

Skin

The skin is the largest organ in the human body. Depending on its function and location, it varies in thickness from 0.5 to approximately 4 mm. For example, around the eyes the skin is the thinnest, and at the base of the feet it is the thickest. Skin thickness can also increase as a result of abrasion and use. This thickening of the skin is mainly restricted to the palms of the hand or the soles of the feet but can also be produced in areas of high abrasive stress. This thickening of the skin is normally referred to as callusing. Figure 7.6 shows a cross section of the main components of the skin.

There are three layers of the skin: the epidermis, the dermis, and the hypodermis. The epidermis is the outermost layer, which is a semipermeable membrane that allows the transfer of fluids out but restricts or acts as a barrier for pathogens. It also holds in water to maintain hydration. Another function of the epidermis is in the regulation of body temperature. The dermis is below the epidermis and acts as a cushioning layer. It is composed of connective tissue that adds elasticity and tensile strength to the skin. The sensory nerve endings are located in the dermis which allow for temperature and tactile sensation. Hair follicles and sweat glands along with blood vessels and other lymphatic vessels and glands are contained in the dermis. The dermis is further divided into the papillary region and the reticular region.

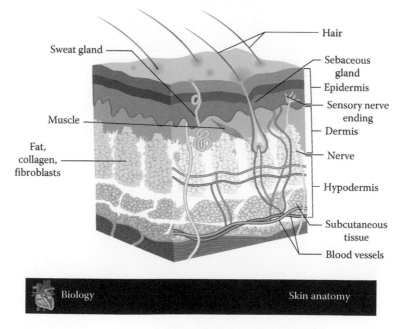

Figure 7.6 Cross section of the main components of the skin.

The papillary region is comprised of connective tissue to the epidermis, whereas the reticular region allows for the elastic and tensile properties of the skin.

The hypodermis is also called the subcutis and is not really part of the skin but rather aids in the attachment of the skin to bones and muscles. This portion is composed of about 50% fat and provides the cushioning effect that protects the underlying structures of the body.

Strength of Human Biological Materials

8

As we see in Chapter 9, biological materials such as bones, muscles, tendons, ligaments, and cartilage are all subject to materials properties. These properties include material properties under tension, compression, bending, impact bending, impact snapping, torsion, tearing, cleavage, shearing, crushing, as well as expansion, bursting, and extraction. Tests yield stress–strain curves under the different tests determining proportional limits, elastic limits, Young's moduli, ultimate limits, creep limits, energy, penetration, and loading effects. Where applicable, we have also included tests on various animals for comparison purposes and to show the similarity in all biological materials.

Most of the data in this chapter is a compilation of tests performed by Yamada and his colleagues over a 25-year period at the Kyoto Prefectural University of Medicine in Kyoto, Japan and detailed in *Strength of Biological Materials*. Most of the tests involved materials from human cadavers and animals. Other data have been culled from a variety of sources, including tests performed by the Society of Automotive Engineers, Wayne State University, and the United States Air Force in preparation for the Space Program in the 1950s and 1960s. Some of the data were also developed in Germany before and during the Second World War. The study and understanding of the structural elements and characteristics of biological materials is paramount in performing calculations relative to how injuries occur and the yield stresses associated with injury. As will be fully explained in Chapter 9, the strength of all materials is based on testing performed over a wide range of forces that produce stress and strain of the material. These tests yield what is commonly referred to as stress–strain curves for the material. Figure 8.1 shows a stress–strain curve that has been generalized to represent some of the points of interest on the curve.

This generalized curve shows the regions that are of interest when assessing injury to humans. That is, certain limits must be exceeded in order to create injury. For example, in order to injure a ligament, its elastic limit must be exceeded. To break a bone, the forces have to exceed the yield point in order to cause a fracture to begin to propagate. Surely, if the ultimate or breaking strength is exceeded, then a catastrophic break occurs and may be typified by a compound fracture. A review of the strength of materials

101

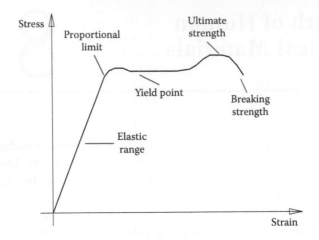

Figure 8.1 Stress–strain.

tables in this chapter for all the biologics shows a range of values not only for certain individuals but also including age-specific data where applicable.

The equipment used to perform these mechanical tests on biological materials included standard mechanical testing equipment such as calipers, micrometers, tension testers, stress–strain recorders, compression testers, Charpy impact testers, Izod impact testers, torsion testers, and Rockwell and Vickers hardness testing apparatus. It is important to note that there may be differences in the values obtained in these tests with in situ values of an actual event. These differences may be especially true for values obtained on soft-tissue specimens in which the test values may vary between the time of death and the time when the specimen is tested. The researchers who have performed these tests have attempted to stabilize the samples in saline solutions and recognized that the strength of these samples will be different than the values obtained soon after death. Consequently for all biological materials, the authors always vary the mechanical properties beyond the values obtained from testing. For example, the lowest value of tensile strength of human intervertebral discs of humans between the ages of 20 and 39 is 0.24 kg/mm². The highest recorded value is 0.33 kg/mm². We would normally use a range of values between 0.20 and 0.30 kg/mm². This range of the variability for the tensile strength allows us to always err on the side of caution or, to put it in another way, to perform a worst-case analysis. By performing a variability analysis for all calculations, the forensic biomechanical engineer allows for any error or unknown that may be present in the calculations. This type of analysis also precludes any questioning that may arise in a deposition or at trial on the validity of the calculations.

The reader may wonder why we would recommend that lower values would be used for the range of a parameter instead of a higher range or a more inclusive range below and above what is published. The reason is as follows: when injury analysis is performed, we look to see if the forces, accelerations, and energies calculated are within the range that would produce the particular effect. Although it is possible, and quite likely, that many individuals have stresses and strengths above the published values, the forces, accelerations, and energies required to produce injury in those individuals is greater. However, if the injury did occur, the more likely cause is a result of a weaker strength of the individual rather than a robust set of parameters. Additionally, an opposing attorney or expert may conclude that the injury could be produced below the published tolerance of a particular biological material. Some experts claim that injuries occur no matter what the stresses applied to the individual are. They state that injuries occur under any and all circumstances. Such claims are a result of conjecture, smoke, and mirrors, and not within the purview of the scientific method. The physical evidence for such claims is scant at best, and generally resulting from allegations by the involved parties. Simply to state that a particular event caused injury is not sufficient. Experts who corroborate such claims never perform calculations to substantiate the claim, and medical examinations generally do not support the "theory" of the injury. To be sure, the scientific literature is completely inconsistent with such claims. We may choose to call these claims junk science, science fiction, charlatanism, deception, distortion of facts, or even outright falsehoods. Regardless of the nature of these unproven and untested claims, they simply do not belong in the realm of science.

Another novel but ridiculous argument used by some experts is that there are trillions of permutations for an injury event that are not considered when computations are performed for the likelihood of injury. They state that these permutations include age, sex, time of day, month, weather, temperature, and many others. They assign numbers to each permutation such as sex (2), age (75), month (12), temperature (100), and so on. They then erroneously multiply the numbers to obtain the numerous permutations. For the given values, we get a number equal to 180,000. You can easily see that if we include 30 days in a month, 24 h in a day, and 3,600 s in an hour, we obtain 466,560,000,000—almost half a trillion permutations by their analysis. Anyone who has taken a course in probability knows that this analysis is flawed, but juries may not be aware of the deception perpetrated. This type of analysis is simply garbage and not science. By simply multiplying all the numbers together, they assume that all events are independent without consideration whether the events are mutually exclusive or not, whether the probability is conditional, or inverse. Let us just look at the male–female question in terms of science. They would say that the number is two, two possibilities or perturbations, because approximately half of the population is male and the other half is female. So, in their calculation on an event, you get the wrong answer half of the time. Is this correct? Absolutely not. We have data for males, and we have data for females. We can apply the range of values for each according to the particular case.

As the authors have a tendency to digress somewhat, we wish to use an analogy for the claims that there are many perturbations of an injury event that make the calculations unsolvable. Consider the classical three-body problem in physics. There are many ways to propose the problem, which states that a closed-form solution for three bodies in space affected by gravitational forces cannot be solved classically with Newtonian mechanics. However, solutions have been found in perturbation theory, Lagrangian points, and Sandman's theorem to name a few. In fact, sending a satellite to the moon is a three-body problem, which has been solved, tested, and performed with extreme accuracy. One aspect of the three-body problem is that the initial conditions cannot be accurately known to send a spacecraft to the moon, so that the solution is not possible. This is the same type of argument made by some experts with respect to solutions for the potential for injury. We can put upper and lower boundaries on the possible solutions in biomechanics and need not tweak the particular solution. When a satellite is sent to the moon, tweaking needs to be performed or else the spacecraft will miss the target. Another analogy is found in computers that are comprised of electronic circuits and devices such as transistors that are not linear but their effect is relativistic. Are we to say that computers do not work? Within boundaries, transistors are gates that express binary logic to perform mathematical operations. The exact condition of the state of the transistor is not necessary in order to determine if a logical state is achieved.

A careful review of the data in this chapter indicates that wherever difference in sex is apparent, it is included. Age differences are also indicated. Additionally, variability in the data is also apparent. These are the main constituents that add variability to the data, along with stature and weight and should be incorporated into biomechanical calculations.

As we did with the other chapters, we begin our discussion of the strength of these biological materials with the skeletal system and the various bones. The skeleton is generally made up of compact bone. Recall that compact bone has the characteristics that provide strength and behave in a support capacity. There are two basic types of bone in an adult, compact or cortical bone and cancellous bone. At the macroscopic level, the distinction between the two types is in terms of the structure. Compact bone comprises approximately 80% of the human skeleton. It has a high Young's modulus. Cancellous or spongy bone is between 30% and 90% porous, is more elastic, and is generally diffused with bone marrow. It has a much lower Young's modulus. Larger bones, such as the femur, have both types, where the outer layers are cortical and the inner portions cancellous. The vertebra is mostly composed of cancellous bone due to its function. The spinal column not only acts as a support mechanism for the upper structure of the human body but also performs a function as much as a shock absorber does in an automobile. It is supple, movable in all three axes, and capable of supporting great weight by distributing the stresses over the entire column. It is a marvel of biomechanical engineering in its complexity and multiple functionality.

In order to summarize and tabulate the properties of the various biological materials, we wish to define the following abbreviations for compactness:

Ultimate tensile strength (UTS)
Ultimate compressive strength (UCS)
Ultimate bending strength (UBS)
Ultimate torsional strength (UTORS)
Modulus of elasticity in torsion (METOR)
Cleavage strength (CS)

Long Bones

We begin the section on the strength of bones by first considering the bones in tension, then compression, and followed by the bending properties. Then we look at the impact bending properties and the torsional properties. For all these sections, we have attempted to include some animal properties for comparison where the data is available or applicable.

Table 8.1 details the ultimate tensile properties of wet compact human bone with respect to the various major bones of the anatomy.

For comparison, Table 8.2 shows human and various animal bones and the respective UTS for the various bones in a wet condition. The wet condition simulates the strength in situ.

Age has an effect on the strength characteristics of human wet compact bone. Table 8.3 shows the effect of age on the strength characteristics of femoral bone in terms of UTC, UCS, UBS, UTORS, METOR, and CS.

The compressive properties of compact animal bone, along with the modulus of elasticity, bending strength, torsional properties, and cleavage properties are listed in Table 8.4. Note that some of the data are available only for humans and not for animals. The compressive data are only available for the femur.

Table 8.1 UTS of Human Compact Bone (kg/mm²)

Bone	Value	Mean
Femur	12.4 ± 0.11	12.4
Tibia	14.3 ± 0.12	14.3
Fibula	14.9 ± 0.15	
Humerus	12.5 ± 0.08	12.5
Radius	15.2 ± 0.14	15.2
Ulna	15.1 ± 0.15	
Average	14	13.6

Table 8.2 Comparison of Human and Animal Wet Compact Bone UTS (kg/mm²)

Humans	Bone	Horses	Cattle	Wild Boar	Pigs	Deer
12.4	Femur	12.1	11.3	10	8.8	10.3
14.3	Tibia	11.3	13.2	11.8	10.8	12
12.5	Humerus	10.2	10.1	10.2	8.8	10.5
15.2	Radius	12	13.5	12.1	10	12.5
13.6	Average	11.4	12	11	9.6	11.3

Table 8.3 Effect of Age on Human Wet Femoral Compact Bone (kg/mm²)

	Ages						
Property	10–19	20–29	30–39	40–49	50–59	60–69	70–79
UTS	11.6	12.5	12.2	11.4	9.5	8.8	8.8
UCS		17	17	16.4	15.8	14.8	
UBS	15.4	17.7	17.7	16.5	15.7	14.2	14.2
UTORS		5.82	5.82	5.37	5.37	4.96	4.96
METOR		350	350	320	320	300	300
CS		9	8.8	8.6		8.2	8.2

Table 8.4 Compressive Properties of Compact Animal Bone (kg/mm²)

Humans	Bone	Horses	Cattle	Wild Boar	Pigs	Deer
16.2	Femur	14.5	14.7	11.8	10	13.3
Modulus of elasticity (kg/mm²)						
320	Femur	940	870	600	490	720
Bending strength (kg/mm²)						
16	Femur	19	18.7			
Torsion (kg/mm²)						
5.41	Femur	5.11				
Cleavage (kg/mm²)						
8.6	Femur	8.8	8.7			

Note that in Table 8.4 the properties of humans, horses, and cattle are essentially the same except for the modulus of elasticity. The shearing properties of human compact wet bone perpendicular to the long axis are detailed in Table 8.5. These properties are also compared with horses, cattle, wild boar, and pigs for the various bones.

Hardness properties may be determined by a variety of methods. The Rockwell Number is a well-known and common method. Table 8.6 includes the hardness of wet and dry human femoral bone by age.

The hardness of animal bone both wet and dry by species is shown in Table 8.7.

Table 8.5   Shearing Properties of Animal Bone Perpendicular to the Long Axis (kg/mm²)

Humans	Bone	Horses	Cattle	Wild Boar	Pigs
8.4	Femur	9.9	9.1	8.7	6.5
8.2	Tibia	8.9	9.5	9	7.1
8.2	Fibula		7.6	8.7	6.2
7.5	Humerus	9	8.6	7.8	5.9
7.2	Radius	9.4	9.3	8.1	6.4
8.3	Ulna		8.2	7.4	4.8
7.9	Average	9.3	8.7	8.3	6.2

Table 8.6   Hardness of Human Femoral Wet Bone Rockwell Number by Age

			Age				
10–19	20–29	30–39	40–49	50–59	60–69	70–79	Average
	49	45	43	39	34	32	40

Dry bone (transverse cross section)

60	66	64	62	58	54	54	60

Table 8.7   Hardness of Animal Bone by Species (Rockwell Number)

Condition	Horses	Cattle	Wild Boar	Pigs	Deer
Wet	76	75	72	55	73
Dry	87	86	82	70	

Table 8.8   Breaking Load in (kg) of Wet Femurs by Age and Sex

		Age		
Sex	20–39	40–59	60–89	Average
Male	5050	4780	4290	4710
Female	4190	3980	3540	3900
Average	4620	4380	3915	4305

Table 8.8 shows the compressive properties of the long bones, in particular the femoral shaft in the longitudinal direction of the middle portion for wet breaking loads for males and females by age group.

Table 8.9 shows the compressive breaking load of portions of the shaft of wet long bones in the longitudinal direction by age, bone, and section of the bone for both males and females.

The UCS of the middle portion of the shaft of wet femoral bone in the longitudinal direction by age is detailed in Table 8.10. Note the slight decrease in strength as the aging process takes place.

Table 8.9 Compressive Breaking Load (kg) of Portions of the Shaft of Wet Long Bones for Ages 20–39 in the Longitudinal Direction

Sex	Tibia	Fibula	Humerus	Radius	Ulna
Third portion					
Male	3660	860	2580	950	1140
Female	2820	590	2100	780	850
Average	3240	725	2340	865	995
Fourth portion					
Male		890	2840	980	
Female		610	2310	800	
Average		750	2575	890	

Table 8.10 UCS (kg/mm²) of the Middle Portion of Wet Femoral Bone in Longitudinal Direction by Age

Age	20–39	40–59	60–89	Average
	15.7	14.9	13.3	14.6

Table 8.11 shows the variation of the UCS of the middle portion of wet bones for humans in the age range of 20–39 in the longitudinal direction for various long bones.

For comparison purposes, Table 8.12 shows the compressive breaking load of the middle portion of animal wet long bones in the longitudinal direction by bone and by animal.

Table 8.13 details the cross-sectional area of the middle portion of the shaft of human wet long bones by sex.

We now turn our attention to the bending and torsional properties of the long bones. As before, these bones are wet to simulate the in situ behavior of bones. The bending properties of human long bones were mostly reported in 1960 by Motoshima. These properties are in the anteroposterior direction by

Table 8.11 UCS (kg/mm²) of the Middle Portion of Wet Bone in the Longitudinal Direction

Femur	Tibia	Fibula	Humerus	Radius	Ulna
15.7	14.2	12.5	12.8	11.7	12

Table 8.12 Compressive Breaking Load (kg) of the Middle Portion of Animal Wet Long Bones in the Longitudinal Direction

Animal	Femur	Tibia	Fibula	Humerus	Radius
Horse	9400 ± 1148				7550 ± 279
Cattle	7134 ± 415				5584 ± 360
Wild Boar	1391 ± 197	980			1390 ± 66
Pigs	1667 ± 130	1250			981 ± 50

Table 8.13 Effective Cross-Sectional Area (mm²) of the Middle Portion of the Shaft of Human Wet Long Bones by Sex and Type

Sex	Femur	Tibia	Fibula	Humerus	Radius	Ulna
Male	330	240	70	200	80	90
Female	260	180	50	160	70	70

Table 8.14 Bending Breaking Load (kg) of Human Wet Bones in Anteroposterior Direction by Age

Age Group	20–39	40–49	50–59	60–69	70–89	Average
Femur	277 ± 11	252 ± 5	240 ± 9	238 ± 6	218 ± 11	250
Tibia	296 ± 1	257 ± 11	248 ± 5	244 ± 9	234 ± 9	262
Fibula	45 ± 2	41 ± 4	40 ± 3	38 ± 2	34 ± 2	40
Humerus	151 ± 12	142 ± 10	131 ± 10	125 ± 9	115 ± 8	136
Radius	60 ± 7	54 ± 4	53 ± 8	49 ± 4	44 ± 3	53
Ulna	72 ± 5	64 ± 8	62 ± 6	60 ± 4	56 ± 4	64

Table 8.15 UBS (kg/mm²) of Human Wet Bones in Anteroposterior Direction by Age

Age Group	20–39	40–49	50–59	60–69	70–89	Average
Femur	12.3 ± 0.34	11.4 ± 0.30	10.6 ±0.23	10.2 ± 0.16	9.6 ± 0.44	11.1
Tibia	10.0 ± 0.42	9.2 ± 0.33	8.6 ± 0.37	8.4 ± 0.21	7.8 ± 0.25	9
Fibula	16.2 ± 0.56	14.6 ± 0.48	13.9 ± 0.39	13.3 ± 0.32	11.8 ± 0.47	14.3
Humerus	10.0 ± 0.15	9.4 ± 0.28	8.3 ± 0.26	7.7 ± 0.21	7.3 ± 0.34	8.8
Radius	10.4 ± 0.24	9.6 ± 0.24	8.9 ± 0.49	8.3 ± 0.22	8.0 ± 0.25	9.3
Ulna	11.1 ± 0.13	10.1 ± 0.31	9.2 ± 0.42	8.4 ± 0.24	8.1 ± 0.46	9.4

age group. Table 8.14 lists the bending breaking load while Table 8.15 lists the UBS of various bones by age group. It was shown that the elastic limit of human wet bones is approximately 70% of the ultimate strength. For the 40–49 age group, the elastic recovery was between 88% and 93%. The creep limit was found to correspond to about 55% of the ultimate strength. The creep limit is the slow deformation of the material under cyclic stress below the yield strength, which eventually exceeds the yield strength and causes failure. No correspondence was found in the bending breaking load of animals. However, the UBS for some animals was determined to be within the range for humans with a much wider variability. Consequently, these data are not included since direct comparisons should not be made nor are they recommended by the authors.

Limited data are available for the impact bending properties of human long bones. For the femur of humans in the age group of 60–79 years, the impact breaking energy was found to be approximately 411 ± 18 kg cm. The estimated value for the age group between 20 and 39 years was determined to be approximately 500 kg cm.

Table 8.16   Torsional Properties of Wet Long Bones of Adult Humans

Average	Femur	Tibia	Fibula	Humerus	Radius	Ulna
Torsional breaking moment (kg cm)						
	1400	1000	116	606	208	190
UTS (kg/mm²)						
4.48	4.62	4.43	4.01	4.35	4.95	4.55
Ultimate angle of twist (°)						
	1.5	3.4	35.7	5.9	15.4	15.2

As with the impact bending properties, the torsional properties of human long bones is limited. Hazama reported the torsional properties of the femur of five subjects while Sonoda et al. reported the properties for other bones. Table 8.16 summarizes those results.

Pertinent stress–strain curves in tension, compression, bending, and torsion for wet compact long bones are included in Figures 8.2 through 8.5 for persons 20–39 years. Figure 8.6 shows compression curves for people 40–59 years.

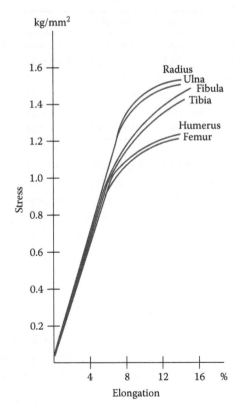

Figure 8.2 Stress–strain in tension of bone for persons 20–39 years.

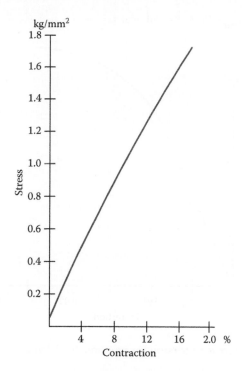

Figure 8.3 Stress–strain in compression of bone for persons 20–39 years.

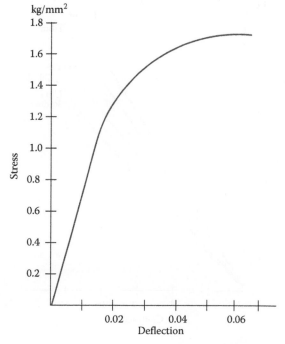

Figure 8.4 Femur stress in bending for persons 20–39 years.

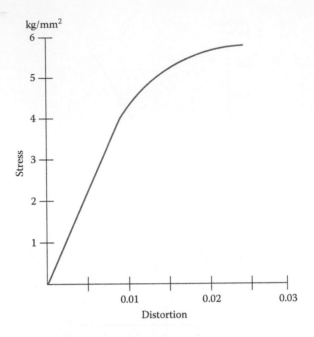

Figure 8.5 Femur stress in torsion for persons 20–39 years.

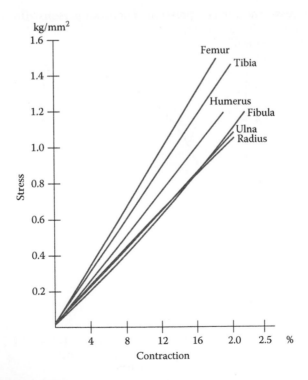

Figure 8.6 Stress–strain in compression for persons 40–59 years.

Spongy Bone

As a way of introduction to vertebrae, we list some basic properties of thoracic and lumbar vertebrae, both in tension and in compression. Table 8.17 shows these values while Figure 8.7 shows the stress–strain curves for spongy bone as a percentage of both in tension and compression.

Vertebrae

As we have stressed throughout the book, one of the most common forms of injury to humans involves the spinal column. These injuries may occur as a result of accidents involving vehicles of all types, lifting movements, rotation, or direct impact, as well as falls. Breaks of vertebra can be catastrophic leading to paralysis or in some rare instances not lead to debilitating conditions. As long as the spinal columns and the branches are not affected,

Table 8.17 Tensile and Compressive Properties of Human Spongy Bone

UTS (kg/mm²) 30–39 years old	Percent elongation 30–39 years old	Average modulus of elasticity
0.12 ± 0.01	0.58% ± 0.03%	8
UCS (kg/mm²) 40–49 years old	Ultimate percent contraction 40–49 years old	Modulus of elasticity (kg/mm²) 40–49 years old
0.19	2.5	9
UCS (kg/mm²) 60–69 years old	Ultimate percent contraction 60–69 years old	Modulus of elasticity (kg/mm²) 60–69 years old
0.14	2.4	7

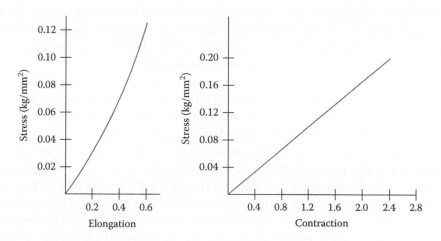

Figure 8.7 Stress–strain wet spongy bone.

Table 8.18   Tensile Properties of Human Wet Vertebrae by Age and Section

Section	20–39 Years	40–79 Years	Average
Tensile breaking load (kg)			
Cervical	114	91	102
Upper thoracic	173 ± 18.9	134 ± 17.9	147
Lower thoracic	336 ± 13.2	279 ± 17.4	298
Lumbar	464 ± 16.7	382 ± 14.5	409
UTS (kg/mm²)			
Cervical	0.35	0.31	0.32
Upper thoracic	0.37 ± 0.01	0.33 ± 0.02	0.34
Lower thoracic	0.38 ± 0.03	0.35 ± 0.02	0.36
Lumbar	0.40 ± 0.03	0.37 ± 0.02	0.38
Average	0.38	0.34	0.35
Ultimate percent elongation			
Cervical	0.85	0.7	0.75
Upper thoracic	0.86 ± 0.11	0.73 ± 0.08	0.77
Lower thoracic	0.87 ± 0.07	0.73 ± 0.09	0.78
Lumbar	0.91 ± 0.13	0.75 ± 0.08	0.8
Average	0.87	0.73	0.78

serious injury may be avoided. In this section, we place particular attention to the tensile, compressive, and torsional properties of human wet vertebrae. Table 8.18 lists the tensile properties of human wet vertebrae in the longitudinal direction by age group and by section. The cervical values are estimated according to Sonoda.

Analysis of the values in Table 8.18 indicates that the lumbar vertebrae are the strongest in terms of breaking load, tensile strength, and resisting the greatest elongation. They are, therefore, the least likely to be injured for a given loading condition. Conversely, the cervical vertebrae appear to be the weakest. The elastic limit is approximately 15% of the ultimate strength, and the creep limit is about 40% of the ultimate strength.

Table 8.19 shows the compressive properties of wet human vertebrae by age and section in the longitudinal direction. Sonoda reported that in the sagittal (anteroposterior) direction, the breaking load of cervical and thoracic vertebrae is 40% of the breaking load of lumbar vertebrae and 60% of the breaking load in the longitudinal (superior–inferior) direction. The breaking load for females is approximately 83% that of males. In the transverse direction, the breaking load of cervical and thoracic vertebrae is 25% of that of lumbar vertebrae and 50% of the breaking load in the longitudinal direction.

No data is available for the torsional properties of the cervical vertebrae. However, these can be gleaned from the torsional properties of the other sections of the spinal columns as detailed in Table 8.20. There are no

Table 8.19   Compressive Properties of Human Wet Vertebrae by Age and Section

Section	20–39 Years	40–59 Years	60–79 Years	Average
Compressive breaking load (kg)				
Cervical	418 ± 6.0	337 ± 9.0	190 ± 6.0	315
Upper thoracic	370 ± 9.0	320 ± 11.3	236 ± 5.6	308
Middle thoracic	431 ± 5.0	373 ± 14.9	232 ± 7.6	345
Lower thoracic	644 ± 24.1	461 ± 24.8	269 ± 7.1	458
Lumbar	730 ± 13.7	477 ± 21.8	308 ± 9.3	505
UCS (kg/mm²)				
Cervical	1.27 ± 0.02	1.08 ± 0.02	0.74 ± 0.02	1.03
Upper thoracic	0.88 ± 0.02	0.73 ± 0.02	0.55 ± 0.02	0.72
Middle thoracic	0.78 ± 0.01	0.62 ± 0.02	0.44 ± 0.01	0.61
Lower thoracic	0.73 ± 0.01	0.53 ± 0.02	0.37 ± 0.01	0.54
Lumbar	0.64 ± 0.01	0.45 ± 0.02	0.31 ± 0.01	0.47
Ultimate percent contraction				
Cervical	8.1 ± 0.08	6.7 ± 0.15	5.1 ± 0.15	6.6
Upper thoracic	7.4 ± 0.15	5.3 ± 0.12	3.8 ± 0.12	5.5
Middle thoracic	6.6 ± 0.11	5.3 ± 0.18	3.6 ± 0.11	5.2
Lower thoracic	5.6 ± 0.09	4.5 ± 0.11	3.5 ± 0.09	4.5
Lumbar	5.6 ± 0.08	4.2 ± 0.09	3.2 ± 0.11	4.3

Table 8.20   Torsional Properties of Human Wet Vertebrae by Age and Section

Section	20–39 Years	40–59 Years	60–79 Years	Average
Torsional breaking moment (kg cm)				
Upper thoracic	60 ± 5.4	48 ± 3.1	40 ± 2.5	49
Middle thoracic	108 ± 7.3	91 ± 4.8	71 ± 4.3	90
Lower thoracic	165 ± 10.2	139 ± 11.1	114 ± 8.6	239
Lumbar	255 ± 18.4	212 ± 14.6	175 ± 9.3	214
UTS (kg/mm²)				
Upper thoracic	0.37 ± 0.04	0.33 ± 0.02	0.30 ± 0.02	0.33
Middle thoracic	0.35 ± 0.01	0.32 ± 0.01	0.29 ± 0.01	0.32
Lower thoracic	0.34 ± 0.02	0.31 ± 0.01	0.27 ± 0.01	0.31
Lumbar	0.32 ± 0.02	0.29 ± 0.01	0.26 ± 0.02	0.29
Ultimate angle of twist (°)				
Upper thoracic	13 ± 0.7	11 ± 0.6	9 ± 0.8	11
Middle thoracic	10 ± 0.8	8 ± 0.7	6 ± 0.6	8
Lower thoracic	8 ± 0.4	6 ± 0.5	5 ± 0.5	6
Lumbar	5 ± 0.6	4 ± 0.5	3 ± 0.5	4

Table 8.21 Height of Human Vertebrae by Age and Section (mm)

Section	20–39 Years	40–59 Years	60–79 Years	Average
Cervical	16.7 ± 0.78	14.6 ± 0.24	11.0 ± 0.11	14.1
Upper thoracic	18.0 ± 0.31	17.5 ± 0.22	14.2 ± 0.27	16.6
Middle thoracic	20.2 ± 0.41	20.0 ± 0.31	18.1 ± 0.36	19.4
Lower thoracic	23.6 ± 0.80	22.3 ± 0.30	21.6 ± 0.22	22.5
Lumbar	27.9 ± 0.43	27.0 ± 0.33	23.6 ± 0.26	26.2

Table 8.22 Cross-Sectional Area of Human Vertebrae by Age and Section (mm²)

Section	20–59 Years	60–79 Years	Average
Cervical	326 ± 7	264 ± 10	305
Upper thoracic	432 ± 13	380 ± 12	415
Middle thoracic	556 ± 18	525 ± 14	546
Lower thoracic	870 ± 34	749 ± 22	830
Lumbar	1088 ± 18	990 ± 21	1055

significant differences in the torsional properties of females as compared to males, except that the breaking moment is approximately 83% for females as those of males. Please refer to Table 8.20 for the torsional properties of human vertebrae.

In order to complete the section on the compressive strength of human vertebrae, we include a table for the height of vertebrae according to age group and section of the spinal column. Table 8.21 details these values and reveals that the lower sections of the spine are larger than the upper portions. All of the tables for the vertebrae and the other bones of the body reveal the effect of the aging process on the human anatomy. Table 8.22 shows the effective cross-sectional area of human vertebrae.

Figure 8.8 shows the stress–strain curves for wet vertebrae in compression and in tension for human adults between the ages of 20 and 39.

Cartilage

Table 8.23 shows the tensile properties of hyaline cartilage as reported by Ko and Takigawa. Table 8.24 shows the compressive properties of hyaline cartilage from studies by Yokoo, Shono, and Asami. Table 8.24 also shows the breaking load for costal cartilage for the corresponding ribs denoted as R2 through R7.

There is no data for the shearing properties of hyaline cartilage for humans. However, in 1964 Ikuki reported the ultimate shearing strength of costal cartilage from horses to be 0.55 ± 0.061 kg/mm² with an ultimate displacement of 2.42 ± 0.12 mm. The proportional limit was found to be

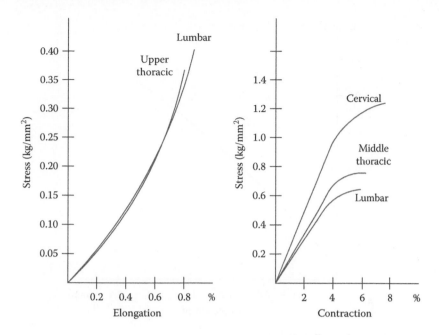

Figure 8.8 Stress–strain vertebrae.

Table 8.23 Tensile Properties of Human Wet Cartilage by Age

0–9 Years	10–19 Years	20–29 Years	30–39 Years	40–49 Years	50–59 Years	60–69 Years	70–79 Years	Average
UTS (kg/mm^2)								
0.46	0.46	0.45	0.42	0.36	0.25	0.15	0.13	0.29
Ultimate percent elongation								
31.2	28.2	25.9	25.5	20.7	16.7	11.2	9.2	18.2

Table 8.24 Compressive Properties of Human Wet Cartilage by Age

10–19 Years	20–29 Years	30–39 Years	40–49 Years	50–59 Years	60–69 Years	Average
UCS (kg/mm^2)						
1.16	0.99	0.96	0.82	0.72	0.72	0.82
Ultimate percent contraction						
17.3	15	13.9	13.9	12.9	12.9	13.6

Compressive Breaking Load of Wet Costal Cartilage (kg)				
Rib No.	10–19 Years	20–39 Years	40–79 Years	Average
R1–R4	67	86	58	67
R5–R7	79	84	62	69
Average	73	85	60	68

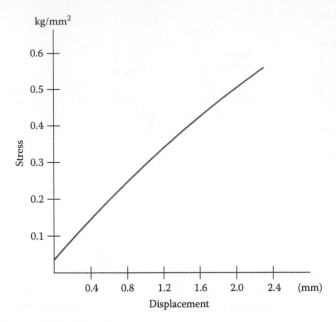

Figure 8.9 Shearing of cartilage.

approximately 65% of the ultimate strength. Figure 8.9 shows the stress–strain curve for the shearing properties of wet hyaline cartilage for horses.

The other two types of cartilage are elastic and fibrocartilage. The tensile properties of human elastic cartilage are displayed in Figure 8.10 for persons in the 40–49-year age range. The UTS in the longitudinal direction was found to be 0.31 kg/mm² for persons in the 40–69-year age range. The percent elongation varied according to the age of the individual. For persons aged 40–49, it was 30%, and for persons aged 60–69, it was 23%.

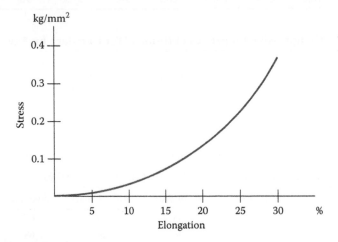

Figure 8.10 Tensile properties of elastic cartilage.

Table 8.25 UTS (kg/mm²) of Fibrocartilage Ages 30–39 Years

| Direction | Layer of Annulus Fibrosis | | |
	Outer	Middle	Inner
Longitudinal	1.60 ± 0.02	1.31 ± 0.02	0.68 ± 0.01
Transverse	0.80 ± 0.01	0.53 ± 0.01	0.45 ± 0.01

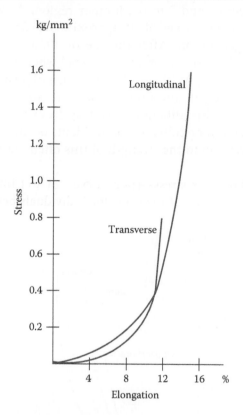

Figure 8.11 Tensile properties of fibrocartilage.

The tensile properties of human fibrocartilage are shown in Table 8.25 and Figure 8.11. These properties were derived by Sonoda from thoracic and lumbar intervertebral discs, both in the longitudinal and transverse directions. These values are for the 30–39-year age group. The elongation limit is approximately 25%–30% of the ultimate strength in both directions.

Discs

One of the most common injuries that is reported or claimed is to the intervertebral discs. The injuries claimed consist of bulges or actual bursting of the outer covering producing a leakage of the gelatinous interior of the disc. The discs, sometimes referred to as intervertebral fibrocartilage, are located between the vertebrae. Essentially, each disc forms a joint that allows a certain amount of movement of the spine that has a dual function to hold the vertebrae together and act as a shock absorber. The annulus fibrosus is the outer portion that surrounds the nucleus pulposus, or gel-like center. There is a total of 23 discs with 6 in the cervical region, 12 in the thoracic region, and 5 in the lumbar region. As we age, the discs begin to degenerate. By the age of 40, approximately 25% of humans show some form of degeneration. After the age of 40, approximately 60% of humans will reveal degeneration at some level when examined via magnetic resonance imaging. With age, the nucleus pulposus begins to dehydrate, decreases in height, and produces a decreasing stature of humans with advancing age. This affliction, resulting from the aging process, is often blamed when minor collisions or accidents occur. Consequently, we pay particular attention to the strength of this often maligned part of the human anatomy.

Figure 8.12 shows the stress–strain curves for wet intervertebral discs in tension for various parts of the spine for individuals between the ages of

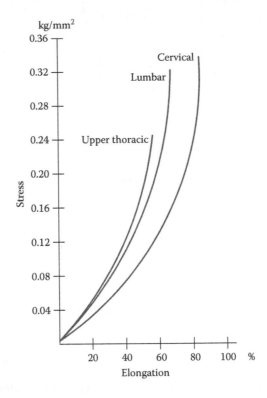

Figure 8.12 Tensile properties of discs.

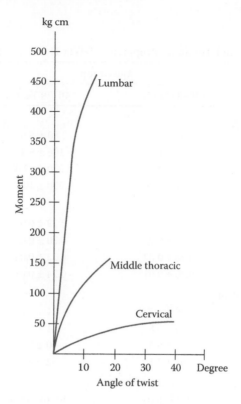

Figure 8.13 Moment angle of twist for discs.

20 and 39 years. Note that the cervical discs exhibit the greatest stress capability simply because they allow for the greatest movement in the neck and head area. Figure 8.13 shows the moment angle of twist for wet intervertebral discs for ages 20–39 for the various sections of the spinal column. Table 8.26 compiles the tensile and torsional properties of intervertebral discs by age group and region of the spine.

Rounding out the section on the strength of discs, Table 8.27 shows the compressive properties of wet intervertebral discs of people in the 40–59-year age range by region of the spinal column. Figure 8.14 shows the stress–strain curve for these data.

Ligaments

There is no comprehensive data on the tensile properties of ligaments of humans. Some values given for cruciate ligaments are in the range of 0.24 kg/mm². However, there are some data on ligaments from cattle.

Table 8.26   Tensile and Torsional Properties of Wet Human Intervertebral Discs

Region	20–39 Years	40–79 Years	Average
	Age Group		
Tensile breaking load (kg)			
Cervical	105 ± 14.5	80 ± 8.6	88
Upper thoracic	142 ± 16.3	106 ± 9.4	118
Lower thoracic	291 ± 21.5	220 ± 12.8	244
Lumbar	394 ± 24.6	290 ± 19.5	325
UTS (kg/mm²)			
Cervical	0.33 ± 0.02	0.29 ± 0.03	0.3
Upper thoracic	0.24 ± 0.01	0.20 ± 0.03	0.21
Lower thoracic	0.26 ± 0.02	0.22 ± 0.01	0.23
Lumbar	0.30 ± 0.01	0.24 ± 0.01	0.26
Ultimate percent elongation			
Cervical	89 ± 4.2	71 ± 3.6	77
Upper thoracic	55 ± 3.8	41 ± 2.1	46
Lower thoracic	57 ± 6.3	40 ± 2.4	46
Lumbar	69 ± 7.1	52 ± 6.2	59
Torsional properties			
	20–39 years	40–69 years	
Torsional breaking moment (kg cm)			
Cervical	56 ± 2.5	48 ± 2.2	51
Upper thoracic	87 ± 3.1	82 ± 4.1	84
Middle thoracic	177 ± 6.0	160 ± 6.3	167
Lower thoracic	273 ± 5.6	260 ± 6.8	265
Lumbar	463 ± 8.9	425 ± 9.7	440
UTS (kg/mm²)			
Cervical	0.52 ± 0.07	0.46 ± 0.05	0.48
Upper thoracic	0.46 ± 0.03	0.38 ± 0.04	0.41
Middle thoracic	0.47 ± 0.02	0.42 ± 0.03	0.44
Lower thoracic	0.48 ± 0.02	0.44 ± 0.04	0.45
Lumbar	0.51 ± 0.03	0.46 ± 0.03	0.48
Ultimate angle of twist (°)			
Cervical	38 ± 5.4	31 ± 2.4	34
Upper thoracic	29 ± 2.2	24 ± 1.6	26
Middle thoracic	24 ± 1.3	20 ± 1.1	22
Lower thoracic	18 ± 1.0	16 ± 0.7	17
Lumbar	15 ± 0.9	13 ± 0.7	14

Table 8.27 Compressive Properties of Wet Intervertebral Human Discs for 40–59 Age Group

		Region		
Cervical	Upper Thoracic	Lower Thoracic	Lumbar	Average
Compressive breaking load (kg)				
320	450	1150	1500	855
UCS (kg/mm²)				
1.08	1.02	1.08	1.12	1.08
Ultimate percent contraction				
35.2	28.6	31.4	35.5	32.67

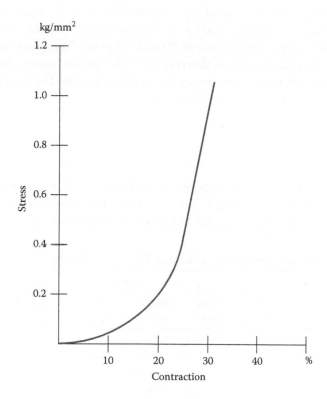

Figure 8.14 Compressive properties of discs.

Sonoda et al. reported in 1962 that the UTS of wet elastic ligamentous tissue from the ligamentum nuchae of cattle is 0.32 ± 0.054 kg/mm² for the restiform portion, 0.20 ± 0.042 kg/mm² for the flat portion, and 0.16 ± 0.052 kg/mm² for the conjunct portion. The thickness of the elastic fiber tested on average to be proportional to the ultimate strength. Recall that restiform means cord-like and conjunct means joined together. The ultimate percent elongation was found to be 160% ± 3.4% in the restiform portion, 128% ± 3.0% in the flat portion, and 99% ± 1.1% in the conjunct portion. The ultimate elongation was found to be proportional to the thickness of the fiber.

Tendons

Tikagawa and Yamada performed extensive studies on the strength characteristics of tendons. Table 8.28 shows the tensile properties of human tendinous tissues by age group and location. Figure 8.15 shows the stress–strain curve for the tensile properties of tendons for persons in the 20–29-year age group.

The estimated tensile breaking load in kilograms of human calcaneal tendons was estimated to be 200 for 20–59 year olds, 190 for 60–69 year olds, and 160 for the age group between 70 and 79 years. The breaking load for female calcaneal tendons was determined to be approximately 80% of that for males. For comparative purposes, the UTS of deer, wild boars, cattle, pigs, and horses varies from 4.2 to 8.4 kg/mm^2.

Muscles

The general consistency of muscles makes them much less strong than the other tissues we have discussed. There are three basic types of muscles: skeletal, cardiac, and smooth. In this book, we are only concerned with skeletal

Table 8.28 Tensile Properties of Human Tendinous Tissue

Tendon	Age Group					
	0–9 Years	10–19 Years	20–29 Years	30–59 Years	60–69 Years	70 to 79 Years
UTS (kg/mm^2)						
Flexor hallucis			6.7 ± 0.20	6.7 ± 0.20		
Extensor hallucis			6.4 ± 0.22	6.4 ± 0.22		
Flexor pollicis			6.4 ± 0.23	6.4 ± 0.23		
Extensor pollicis			6.2 ± 0.22	6.2 ± 0.22		
Calcaneal	5.3	5.6	5.6 ± 0.09	5.6 ± 0.09	5.3 ± 0.14	4.4 ± 0.17
Ultimate percent elongation						
Flexor hallucis			9.4 ± 0.27	9.4 ± 0.27		
Extensor hallucis			9.9 ± 0.28	9.9 ± 0.28		
Flexor pollicis			9.6 ± 0.28	9.6 ± 0.28		
Extensor pollicis			9.9 ± 0.26	9.9 ± 0.26		
Calcaneal	11	10	9.9 ± 0.11	9.5 ± 0.14	9.1 ± 0.23	9.1 ± 0.23

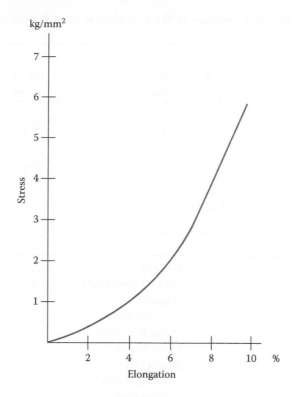

Figure 8.15 Tensile properties of tendons.

muscle, which is susceptible to injury. We are not concerned with cardiac muscle that powers the heart or smooth muscle in the digestive system and various other organs. Both smooth and cardiac muscles are involuntary and necessary for the sustainability of life. In contrast, skeletal muscle is purely voluntary, is anchored by tendons attached to bones, and used for locomotion. The composition of normal humans is approximately 35%–40% skeletal muscle depending on gender with females at the lower percentage. Muscle strength in this section should not be confused with the ability of a particular individual to perform a feat of strength. Remember, we are only concerned with the properties of this type of tissue.

Muscles generally fail in tension, which is often associated with a failure of a tendon or ligament. Table 8.29 shows the tensile properties of the rectus abdominis muscles by age group, while Table 8.30 shows the UTS of various muscles for the 20–39-year age group. Figure 8.16 shows the stress–strain curves for various muscles for persons 29 years of age.

There is no significant difference in the strength of skeletal muscle according to gender. However, it was found that the ultimate strength of the muscles of thin men is approximately 1.8 times that of normal men.

Table 8.29   Tensile Properties of Skeletal Muscle from Rectus Abdominis by Age

			Age Group			
10–19 Years	20–29 Years	30–39 Years	40–49 Years	50–59 Years	60–69 Years	70–79 Years
UTS (g/mm²)						
19 ± 1.2	15 ± 0.6	13 ± 1.0	11 ± 0.6	10 ± 0.5	9 ± 0.3	9 ± 0.3
Ultimate percent elongation						
65 ± 1.2	64 ± 1.1	62 ± 0.7	61 ± 0.9	61 ± 1.5	58 ± 1.8	58 ± 1.8

Table 8.30   UTS (g/mm²) of Skeletal Muscle by Classification for Ages 20–39 Years

Classification	Muscle	UTS
Mastication	Masseter	13
Trunk	Sternocleidomastoid	19
	Trapezius	16
	Pectoralis major	13
	Rectus abdominis	14
Upper extremity	Biceps brachii	17
	Triceps brachii	21
	Flexor carpi radialis	15
	Brachioradialis	18
Lower extremity	Psoas major	12
	Sartorius	30
	Gracilis	20
	Rectus femoris	10
	Vastus medialis	15
	Adductor longus	13
	Semimembranosus	13
	Gastrocnemius	10
	Tibialis anterior	22

No relationship exists between the ultimate strength of muscle tissue and the thickness of the muscle fiber.

Teeth

Injuries to the face may involve the dislocation or the breaking of teeth. Classic examples from sports are hockey players although boxers, basketball players, and other athletes are prone to such injuries. The loss of teeth may also be caused by blunt trauma caused by another individual or in many accidents. A doubles tennis player may get hit with a tennis racket by his

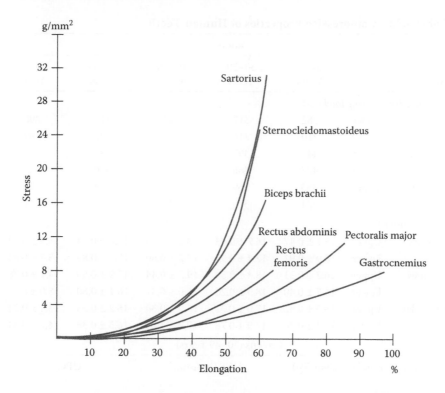

Figure 8.16 Stress–strain curve for muscles.

teammate and lose some teeth. Table 8.31 lists the extractive properties of teeth and the UTS by type and age group as determined by Yoshimatsu et al.

The compressive properties of dental crowns and roots are listed in Table 8.32. For the crowns, the table is divided by location and the occlusal surface of the crown for people from 30 to 49 years of age. For the root, the table lists the type of tooth by age group.

Table 8.31 Extractive Properties of Teeth

	Extractive Load (kg)		UTS (kg/mm²)	
Type	20–29 Years	40–49 Years	20–49 Years	Location
Incisors	24	22	0.15	Upper
	22	17	0.16	Lower
Canines	27		0.17	Upper
	39	32	0.16	Lower
Premolars	30	28	0.14	Upper
	24	24	0.14	Lower
Molars	36	30	0.12	Upper
	36	23	0.12	Lower

Table 8.32 Compressive Properties of Human Teeth

Age		Roots				
		20–29 Years	30–39 Years	40–49 Years	50–59 Years	60–69 Years
Compressive breaking load (kg)						
Incisors	Upper	282	312	342	316	290
	Lower	210	216	214	189	189
Canines	Upper	440	497	525	482	380
	Lower	427	466	475	410	382
Premolars	Upper	353	369	352	332	332
	Lower	274	319	307	290	261
UCS (kg/mm²)						
Incisors	Upper	15.1 ± 0.47	16.7 ± 0.43	18.3 ± 0.66	17.0 ± 0.74	15.5 ± 0.42
	Lower	16.9 ± 0.91	17.4 ± 0.93	17.2 ± 0.60	15.2 ± 0.83	15.2 ± 0.83
Canines	Upper	16.3 ± 0.31	18.4 ± 0.54	19.5 ± 0.44	17.9 ± 0.55	14.1 ± 0.28
	Lower	16.7 ± 0.40	18.3 ± 0.20	18.6 ± 0.31	16.1 ± 0.80	15.0 ± 0.33
Premolars	Upper	16.3 ± 0.82	17.0 ± 0.31	16.2 ± 0.30	15.2 ± 0.55	13.4 ± 0.24
	Lower	15.4 ± 0.36	17.9 ± 0.78	17.2 ± 1.27	16.3 ± 0.38	14.7 ± 0.64

Crowns (30–49 Years)			
Compressive Breaking Load (kg)		UCS (kg/mm)²	UPC
Second premolars			
	Upper	720	14.9	2.28
	Lower			
First molars			
	Upper	977	13.8	2.85
	Lower	937	14.9	2.66
Second molars			
	Upper	1000	15.1	3.1
	Lower			

Skin

Our final section on the strength of biological materials is for the human body's largest organ, the skin. Skin properties vary significantly according to location on the body. In 1960, Yamaguchi tested the tensile properties of a total of 66 individuals ranging in age from 10 to 79 years of age. These tests were performed in both the longitudinal and transverse directions. His findings are summarized in Tables 8.33 and 8.34.

The skin is the first line of defense with respect not only to infection but also to injury. It is a window into the location and to a certain extent the

Table 8.33 Tensile Properties of Human Skin

Region	Tensile Breaking Load per Unit Width (kg/mm)		UTS (kg/mm²)		Ultimate Percent Elongation	
	30–39 Years	Adult Average	30–49 Years	Adult Average	10–29 Years	Adult Average
Head	0.80	0.90	0.40	0.36	57.00	44.00
Forehead	1.00	0.90	0.51	0.46	70.00	54.00
Cheek	1.30	1.10	0.62	0.56	91.00	70.00
Neck	2.20	1.60	1.32	1.19	120.00	93.00
Upper abdomen	2.40	2.10	1.13	1.02	126.00	98.00
Lower abdomen	2.20	2.00	1.16	1.04	123.00	94.00
Nape	3.40	3.10	1.13	1.02	109.00	84.00
Back	3.00	3.00	1.00	0.90	110.00	85.00
Buttock	2.10	1.80	0.77	0.69	117.00	90.00
Arm front	2.00	1.40	1.15	1.04	123.00	95.00
Arm back	2.20	1.70	1.05	0.94	120.00	93.00
Forearm front	1.90	1.30	1.19	1.07	98.00	76.00
Forearm back	2.40	1.50	1.04	0.94	96.00	74.00
Palm	2.50	1.70	0.87	0.78	56.00	43.00
Hand back	1.00	0.70	0.63	0.57	84.00	65.00
Thigh back	2.40	1.80	1.00	0.90	114.00	88.00
Thigh front	2.10	1.50	0.84	0.76	121.00	94.00
Leg back	1.70	1.20	0.80	0.72	111.00	86.00
Leg front	2.20	1.60	0.88	0.79	102.00	79.00
Sole	2.50	1.90	0.95	0.86	62.00	48.00
Foot top	1.70	1.50	0.99	0.89	80.00	62.00

Table 8.34 Effect of Age on Tensile Properties of Skin Based on Maximum Value Ratio

10–29 Years	30–39 Years	50–59 Years	60–69 Years	70–79 Years
Tensile breaking load per unit width				
0.70	1.00	0.78	0.70	0.46
UTS				
0.65	1.00	0.91	0.82	0.70
Ultimate percent elongation				
1.00	0.80	0.73	0.64	0.57

severity of the injury whether superficial or deeply seated. Medical practitioners are trained to detect injury first by noting the condition of the skin and then assessing the types of diagnostics that may follow in order to make an accurate assessment of the extent and type of injury. The data on the properties of skin can be used to assess penetration type injuries from blunt or sharp objects.

Mechanics of Materials 9

Stress and Strain

Analysis of structural elements requires knowledge of the mechanics of materials. These principles permit a determination of stress and strain within a material subject to loading. These loads, or forces, acting on a body may be internal or external. The forces that resist separation or movement within the body are considered internal, while external forces are those transmitted by the body. Generally, the force acting on a body is proportional to the stress times the cross-sectional area:

$$F = \sigma A \qquad (9.1)$$

Stress σ can also be referred to as the intensity of forces distributed over the cross section. Stresses imparted on a body induce deformations, which can change the shape, length, and cross-section of a body. In the design of engineering materials, it is important to avoid large deformations that affect the serviceability of a structure or machine. For the human body, pain and injury can result from excessive deformation under loading.

The deformation δ of a body per unit length L is known as the strain ε:

$$\varepsilon = \frac{\delta}{L} \qquad (9.2)$$

The relationship between stress and strain is best described in a Cartesian diagram, with stress σ as the y-axis and strain ε as the x-axis.

This generalized diagram is not specific to any particular material. It should be noted that all materials behave differently when subjected to loading. Information regarding the response of biological materials to stress and strain can be found in Chapter 8. In this case, the stress–strain curve is linear until the yield point. The yield point is defined by the principle that the material will behave elastically if the stress does not exceed this limit. Elastic behavior means that there is no permanent strain, or deformation, of a material. The relationship between stress and strain before this yield point is commonly referred to as Hooke's law:

$$\sigma = E\varepsilon \qquad (9.3)$$

This relationship is linear for the elastic phase of the stress–strain diagram. The linear coefficient, E, is known as Young's modulus, or the modulus of elasticity. For the example of an axially-loaded bar, strain is the ratio between the change in length (deformation) and the total length. Thus, strain is a dimensionless parameter, which indicates that Young's modulus is to be expressed in units of stress. For the English system of units, stress will often be given in terms of pounds per square inch (psi). Pascal (Pa) will commonly be used when SI units are considered. A pascal is a newton per meter squared, or kilogram per meter per second squared:

$$1\,\text{Pa} = 1\,\text{N/m}^2 = 1\,\text{kg/ms}^2$$

It is often useful to examine the response of materials under loading beyond the yield point. In statics, which is discussed in Chapter 11, there is an underlying assumption that the structures being analyzed are rigid and nondeformable. However, some structures are indeterminate when analyzed statically. Consideration of deformation of structural members permits the computation of forces for statically indeterminate structures. The following are representative stress–strain diagrams for typical ductile and brittle materials (Figure 9.1).

Ductility refers to the ability of a material to deform beyond the yield point. While permanent deformations will be induced once the load is

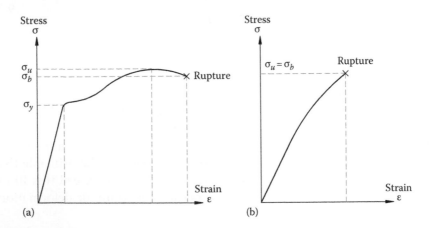

Figure 9.1 Stress–strain curves: (a) ductile material and (b) brittle material.

removed, rupture does not occur as long as the applied stresses remain below an allowable limit, which is referred to as its ultimate strength. The stress–strain diagram for a ductile material will involve a flattening beyond the yield point, wherein the plastic behavior of the material takes hold. Beyond the yield point, removing the load will result in a permanent, plastic deformation of the material. This permanent deformation may be represented by drawing a sloping line from the point of maximum stress back to the x-axis, where stress is equal to zero. The slope of the line is equal to the ratio of stress over strain within the elastic region, which is equal to Young's modulus E.

Brittle materials rupture without any observable change in deformation rates. For these materials, the ultimate strength is equal to its breaking strength. There is much less strain at the time of rupture for a brittle material as compared to a ductile material. There is essentially no plastic response for a brittle material. Thus, permanent strains and deformations will be negligible as long as the loads do not exceed the breaking limit.

In terms of the mechanics of biological materials, specific stress–strain curves have been developed, as outlined in Chapter 8. These curves are based on tests conducted on various elements of a human body, including bones, ligaments, and tendons. The tests involved the variety of loading conditions and types of stresses developed by these materials. A basic description of these stresses will follow.

Axial Stresses: Compression and Tension

Consider a straight, slender rod shown in Figure 9.2. The rod is loaded in the direction of its longitudinal axis. Thus, this member is under axial loading. The axial loads are considered to be in compression if they are directed into the member, whereas tension loads are produced if the forces are pulling on the member.

For an axial member with a nonuniform cross section, Equation 9.1 can be modified as shown in the following equation:

$$F = \int dF = \int_A \sigma \, dA \qquad (9.4)$$

From equilibrium, a force applied to the axial member must have an equal and opposite force. For concentric loading and uniform cross sections, estimates of normal stresses may be made for some two-force members. However, in some situations, the solutions may not be available through statics. A member may be statically indeterminate for a slender rod fixed at both ends and

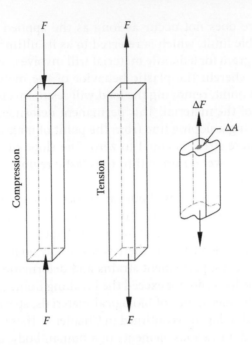

Figure 9.2 Axial loading on a slender member.

placed under an eccentric, axial load at midspan. The solution of the distribution of normal stresses in this member requires additional knowledge of the manner of loading and consideration of deformations, which is discussed in Chapter 11.

Figure 9.3 depicts an axial stress–strain curve of long bones for humans. This curve was developed by Yamada and has been modified to examine the properties of the femur.

Curve fitting permits a determination of the yield stress, ultimate strength, and percent elongation at these points. A cursory examination of these curves reveals that these bones are relatively brittle in tension.

Curve fitting provides an approximate yield stress of 6.0 kg/mm². This value is accompanied by a percent elongation of 0.34%, or a change of 0.0034 mm/mm. The estimate of Young's modulus E is therefore 1765 kg/mm². From the principal data source (Yamada), the mean modulus of elasticity is listed as 1760 kg/mm² for this bone in tension. The curve fitting yields an ultimate strength of 12.26 kg/mm² and ultimate percent elongation of 1.38%. These values are consistent with the mean values reported in Yamada, which are 12.4 kg/mm² and 1.41%, respectively.

With the exception of the yield stress, basic data on ultimate stresses are available in tabular format. Thus, the tables would normally render such curve-fitting methods as duplicative. As seen in Chapter 8, both graphical and tabular formats are provided for most of the biomechanical characteristics.

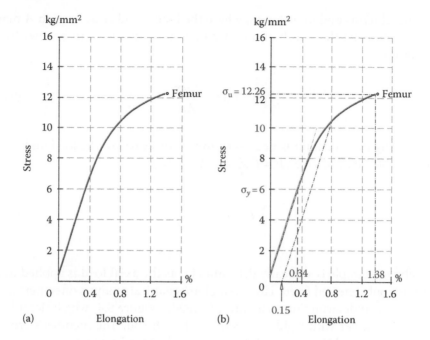

Figure 9.3 (a) Stress–strain curves in tension of wet compact bone from the limb bones of persons 20–39 years of age (Yamada). (b) Stress–strain curves in tension of femur—curve fitting.

However, additional information can be gleaned, especially for material response within the plastic region of the curve. For example, the application of a stress of 10 kg/mm², which exceeds the estimated yield stress of 6.0 kg/mm², results in a permanent elongation of the bone after removing the load. By drawing a line sloped at E down from the intersection of the curve with the maximum applied load (10 kg/mm²), the permanent strain can be estimated. As shown, the bone would be expected to be permanently elongated by 0.15% in this example.

As is common in brittle materials, the ultimate strength in compression is normally much larger than the tensile ultimate strength. This difference may be attributable to microscopic voids or flaws in the bone, which only manifest weaknesses in the material when tension loads are applied. However, the materials behave similarly in tension and compression when stresses are lower than the yield point.

For the case of a bar in tension with one end fixed, Equations 9.1 and 9.2 can be rewritten in terms of the deflection under load:

$$\delta = \frac{FL}{AE} \tag{9.5}$$

In general terms and in instances where the load F and cross section A may be variable along the length of the specimen, this equation can be rewritten as follows:

$$d\delta = \varepsilon\,dx = \frac{F\,dx}{AE} \tag{9.6}$$

Integrating this equation is necessary in instances where the load and area are variable functions (i.e., $F = F(x)$ and $A = A(x)$):

$$\int_{0}^{L} d\delta = \delta = \int_{0}^{L} \frac{F\,dx}{AE} \tag{9.7}$$

The above example is statically determinate, as the axial load is applied to a bar with only one end fixed. However, biomechanical analysis often pertains to statically indeterminate structures. Consider Figure 9.4, which describes an axial member with rigid connections at each end. The member's cross-sectional area and Young's modulus are considered to be constant.

The member is loaded concentrically along its longitudinal axis. Thus, there is no moment acting on the member and the forces act in one direction. By summing forces, the following equation is derived:

$$\sum F_x = 0 : R_1 + P - R_2 = 0 \rightarrow P = R_2 - R_1 \tag{9.8}$$

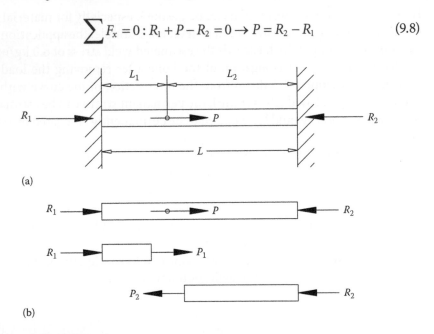

(a)

(b)

Figure 9.4 (a) Axial member with fixed ends and (b) free-body diagram.

Given that there are two unknowns and one equation, this problem is statically indeterminate. However, consideration of the deformations induced by loading allows this problem to be solved.

As both ends are rigid and fixed, it can be stated that the total deformation of the member is zero. Figure 9.4b reflects a free-body diagram, wherein the internal loads on either side of the load point are considered. The total deformation is taken as the summation of the deformation from the internal loads P_1 and P_2:

$$\sum \delta = 0 : \delta_t = \delta_1 + \delta_2 = \frac{P_1 L_1}{AE} + \frac{P_2 L_2}{AE} = 0 \tag{9.9}$$

By reviewing the free-body diagram, it is evident that $P_1 = R_1$ and $P_2 = R_2$. The equation can further be simplified by eliminating the AE variables, as these quantities are constant for this example:

$$\sum \delta = 0 : R_1 L_1 + R_2 L_2 = 0 \tag{9.10}$$

Now we have two equations for two unknowns. The deformation equation can be solved simultaneously with the relation governed by summing forces:

$$R_1 = \frac{-PL_2}{L}; \quad R_2 = \frac{PL_1}{L} \tag{9.11}$$

From the basic relationship between loading and normal stress for a member of constant area, the following stresses are determined:

$$\sigma_1 = \frac{-PL_2}{LA}; \quad \sigma_2 = \frac{PL_1}{LA} \tag{9.12}$$

Shear

The earlier discussion involves the determination of axial stresses, either in compression or in tension. These stresses are normal, or perpendicular, to the considered surface. Now consider conditions in which the applied loads are parallel to the cross section that is being examined. Figure 9.5 demonstrates the orientation of such transverse loads on a structural element.

Here, the load P is equal to the shear V across the section. The shear stress is defined by the Greek symbol τ. A generalized distribution of

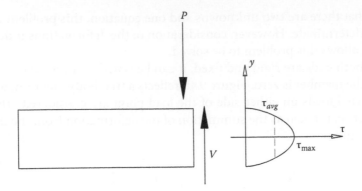

Figure 9.5 Transverse loading.

shearing stress across the section is also shown. For a rectangular shape, this distribution is parabolic and the maximum stress (τ_{max}) occurs along the centroidal axis. The shearing stresses are zero at the edges of the member. Depending on the sectional parameters of the member, an average shearing stress (τ_{avg}) may be computed. The average shearing stress is given by the following:

$$\tau_{avg} = \frac{V}{A} = \frac{P}{A} \tag{9.13}$$

Note that this average value should not be assumed to be uniform across the section, which is in contrast to the distribution of normal stresses. In structural design, it is often useful to determine the maximum possible stress. Similarly, the results of biomechanical analyses may depend on the maximum shearing stress, which requires knowledge of the stress distribution, and not merely the average value.

Shearing is a controlling limit in the structural design of connections. It follows that shearing is of particular concern with ligaments, tendons, and other connecting structures in the human anatomy. In instances of bending of a short-span beam with a deep cross section, shearing may also control the design process. However, such structures are not commonly found in the human body.

Depending on the direction of loading, stresses developed at ball and socket joints may be referred to as bearing stresses. Examples of these joints are found where the femoral head fits within the hip socket and at the connection between the humerus and shoulder blade. The distribution of bearing stresses is not uniform, and the average value is normally considered. For a curved bearing surface, the bearing stress is determined by the ratio of the load P to the projected area of the connection surface. This projected area may be considered as the width b of the bearing surface times the thickness t of the area of contact:

$$\sigma_b = \frac{P}{A} = \frac{P}{bt} \tag{9.14}$$

Oblique Loading

In the previous examples, we examined the distribution of stresses for axial and shear loading. In both cases, the stresses were determined along planes perpendicular to the axis of the member. However, if the considered plane is not orthogonal to the main axis, both axial and transverse loading will result in the development of normal and shearing stresses across the nonorthogonal planes as shown in Figure 9.6.

This manner of loading produces a normal load (P_n) and transverse load (P_t) across the oblique surface. The terms A_o and A_t refer to the effective area of the normal and oblique surfaces, respectively. These variables are defined as follows:

$$P_n = P\sin\Theta \quad \text{and} \quad P_t = P\sin\Theta$$

$$A_o = A_t \sin\Theta \Rightarrow A_t = \frac{A_o}{\sin\Theta} \tag{9.15}$$

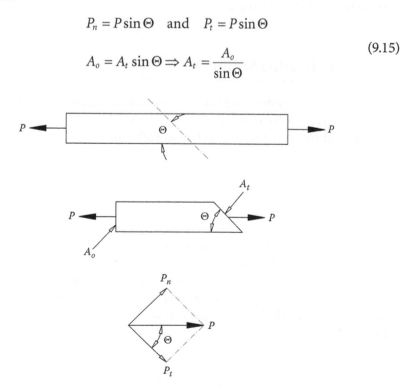

Figure 9.6 Oblique loading.

The normal stress across the oblique surface is equal to the normal load divided by the oblique surface area:

$$\sigma = \frac{P_n}{A_t} = \frac{P\sin\Theta}{A_o/\sin\Theta} = \frac{P}{A_o}\sin^2\Theta \qquad (9.16)$$

The average shearing stress is given by the ratio of the transverse load to the oblique surface area:

$$\tau = \frac{P_t}{A_t} = \frac{P\cos\Theta}{A_o/\sin\Theta} = \frac{P}{A_o}\sin\Theta\cos\Theta \qquad (9.17)$$

If $\Theta = 90°$, the surface is orthogonal and nonoblique. Additionally, the sine of this angle is unity, and the cosine is zero. Thus, the normal stress is equal to the load divided by the normal surface area, and the shear stress is zero. It can be shown that the shearing stress reaches its maximum value for a surface inclined at a 45° angle.

Axial and Shearing Strain

Consider again a uniform and homogeneous bar loaded about its longitudinal axis and represented by Figure 9.7.

The axial strain (ε_x) is defined by Hooke's law:

$$\varepsilon_x = \frac{\sigma_x}{E} \qquad (9.18)$$

It may be assumed that the lateral strain (ε_y) would be zero, as there is no applied load in the y-direction. However, such an assumption would be erroneous. As shown in Figure 9.7, the load P induces elongation of the bar in the longitudinal (x) direction, which in turn contracts the member in the transverse (y) direction. The lateral strain is related to the axial strain by Poisson's ratio, ν:

$$\nu = \left|\frac{\text{lateral strain}}{\text{axial strain}}\right| \qquad (9.19)$$

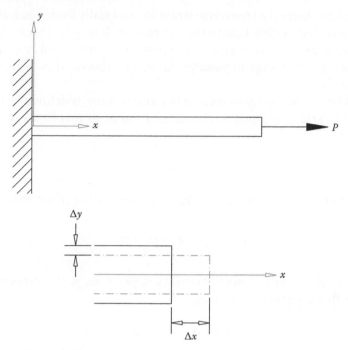

Figure 9.7 Axial loading.

In the example shown, the elongation in the positive x-direction is accompanied by a contraction in the negative y-direction. Thus, Poisson's ratio can be written as:

$$v = \frac{-\varepsilon_y}{\varepsilon_x} \tag{9.20}$$

Furthermore, this ratio can provide a value for lateral strain in terms of normal stress and the modulus of elasticity:

$$\varepsilon_y = -v\varepsilon_x = -\frac{v\sigma_x}{E} \tag{9.21}$$

The shearing strain (γ) is related to shearing stress (τ) by the modulus of rigidity, G. This value may also be referred to as the shear modulus. For a 3D object, the shearing stresses about the x–y, y–z, and z–x planes are as follows:

$$\tau_{xy} = G\gamma_{xy}$$

$$\tau_{yz} = G\gamma_{yz} \tag{9.22}$$

$$\tau_{zx} = G\gamma_{zx}$$

As shown here, there is a transverse strain for an axially loaded bar, although there is no corresponding load in the y-direction. It may further be assumed that the combined effects of axial and transverse strain would be equivalent, thus negating any change in volume. As will be shown, this assumption is also erroneous.

The dilation, or change in volume per unit volume, is defined by the variable e. This variable is equal to the summation of strains in the x-, y-, and z-directions:

$$e = \varepsilon_x + \varepsilon_y + \varepsilon_z \tag{9.23}$$

This equation can be rearranged to describe the dilation in terms of stress:

$$e = \frac{1-2v}{E}(\sigma_x + \sigma_y + \sigma_z) \tag{9.24}$$

For a body subject to a uniform hydrostatic pressure, p, the stress variables can be further simplified as follows:

$$-p = \sigma_x = \sigma_y = \sigma_z \tag{9.25}$$

The dilation term can therefore be described as follows:

$$e = -\frac{3(1-2v)}{E}p = -\frac{p}{k}$$
$$\text{where } k = \frac{E}{3(1-2v)} \tag{9.26}$$

The k term is defined as the bulk modulus, or modulus of compression.

A material subject to hydrostatic pressure can only decrease in volume. Therefore, the dilation e should be less than zero, which in turn means that the bulk modulus k should be greater than zero. Furthermore, if k is greater than zero, then the following can be said about Poisson's ratio:

$$k = \frac{E}{3(1-2v)} > 0 \Rightarrow (1-2v) > 0 \Rightarrow v < \frac{1}{2} \tag{9.27}$$

As shown, Poisson's ratio v cannot be greater than ½. From its definition, Poisson's ratio is the absolute value of the ratio of lateral strain to axial strain. Thus, this ratio cannot be less than zero. In other words,

$$0 < v < \frac{1}{2} \tag{9.28}$$

For a maximum Poisson's ratio of ½, the following can be determined:

$$v = \frac{1}{2} \Rightarrow k = \frac{E}{3(1-2(1/2))} = \frac{E}{3(0)} = \infty \Rightarrow e = -\frac{p}{\infty} = 0 \qquad (9.29)$$

As shown, the material has an infinite bulk modulus and a dilation equal to zero. Such a material would be considered to be incompressible. No material is completely incompressible, and certainly no biological material will have a Poisson's ratio of ½.

Recall the equation for dilation in terms of stress. For an axially loaded bar, the stress in the x-direction (σ_x) is greater than zero, while the remaining stresses (σ_y and σ_z) are zero:

$$e = \frac{1-2v}{E}(\sigma_x + \sigma_y + \sigma_z) = \frac{1-2v}{E}(\sigma_x) \qquad (9.30)$$

Since Poisson's ratio is less than ½, the dilation is greater than zero. Thus, there is a positive change in volume for an axially loaded bar.

Torsion

In this section, we analyze stresses and strains in members subject to a torque T. Torque is a twisting couple, or moment, applied to a structural member. For the purposes of deriving governing equations for torsion, the members are assumed to be circular in cross section. No material within the human body is perfectly, or even nearly, circular. However, an idealized model may be used for some biomechanical applications. A discussion of empirical solutions for torsion of noncircular members will be presented as well.

Figure 9.8 details a circular bar of length l and an applied torque T.

The top diagram details the angle of twist φ of a circular bar fixed at one end. As will be shown, this angle is proportional to the applied torque T and length l of the bar. Thus, a bar twice as long as another bar, both subject to the same torque and possessing the same cross section, will endure double the angle of twist. In circular shafts, it is assumed that every cross section remains planar and undistorted when subject to torque. This principle applies because circular bars are axisymmetric. The same cannot be said for elements of the human anatomy.

In the bottom diagram of Figure 9.8, ρ is the radial distance of the differential area dA from the centroidal axis. The differential force dF acts

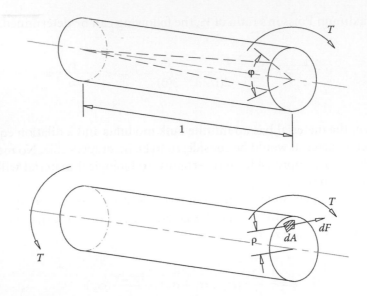

Figure 9.8 Torsion of a circular bar.

perpendicular to the radial distance. The torque is determined by summing the product of these forces and distances:

$$T = \int \rho \, dF \qquad (9.31)$$

From the earlier sections, the average shearing stress is the ratio of force per unit area. Therefore,

$$dF = \tau \, dA \qquad (9.32)$$

Combining these two equations yields the following relationship:

$$T = \int \rho \tau \, dA \qquad (9.33)$$

The shearing stress is zero at the center axis and is greatest in magnitude along the perimeter of the circular shaft. Integration yields the shearing stress produced by an applied torque:

$$\tau = \frac{T\rho}{J} \qquad (9.34)$$

where

ρ is the radial distance from the centroidal axis

J is the polar moment of inertia

For a circle of radius r, the polar moment of inertia is given by the following:

$$J = \frac{\pi}{2}r^4 \tag{9.35}$$

For bones, there is a core of marrow that provides no appreciable strength or resistance to torque. For a hollow cylinder with inner radius r_i and outer radius r_o, the polar moment of inertia is as follows:

$$J = \frac{\pi}{2}r_o^4 - \frac{\pi}{2}r_i^4 = \frac{\pi}{2}\left(r_o^4 - r_i^4\right) \tag{9.36}$$

The angle of twist within the elastic range is the product of torque T and length l divided by the polar moment of inertia J and shear modulus G:

$$\tau = \frac{Tl}{JG} \tag{9.37}$$

Consider Figure 9.9, which demonstrates the torsion of a rectangular bar.

From experimentation, there are no deformations, and therefore no stresses, along the edges of the edges. The largest deformations and stresses occur along the center line of each face of the bar. Mathematical theory of elasticity provides equations for bars of constant cross section. For a rectangular bar with a width a and height b, the maximum shearing stress and angle of twist are given by the following:

$$\tau_{max} = \frac{T}{c_1 ab^2}$$

$$\tag{9.38}$$

$$\varphi = \frac{Tl}{c_2 ab^3 G}$$

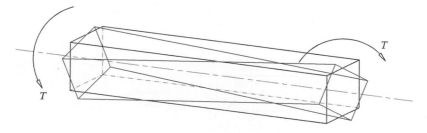

Figure 9.9 Torsion of a noncircular member.

The maximum shearing stress occurs along the center line of the wider face a of the bar. The coefficients c_1 and c_2 are based on the dimensions a and b of the bar. For example, a bar having an $a{:}b$ ratio of 2.0 will have a c_1 coefficient of 0.246 and c_2 coefficient of 0.229. These coefficients converge to the same value (0.333) as the $a{:}b$ ratio approaches infinity.

Bending

Figure 9.10 describes a prismatic beam of length l with applied moments M on each end.

The lower diagram details the stress distribution across the face of the beam. The neutral axis represents the surface parallel to the top and bottom of the beam wherein the normal stresses and strains are zero. The distance c represents the largest distance from the neutral axis to the edge of the cross sections. The y-axis begins at any point along the neutral axis.

Consider Figure 9.11, which demonstrates a prismatic beam bending in a curved shape.

From this diagram, the longitudinal normal and maximum strains are described by the following:

$$\varepsilon_x = -\frac{y}{\rho} \quad \text{and} \quad \varepsilon_{max} = \frac{c}{\rho} \tag{9.39}$$

From these equations, the normal strain can be related to the maximum strain:

$$\varepsilon_x = -\frac{y}{c}\varepsilon_{max} \tag{9.40}$$

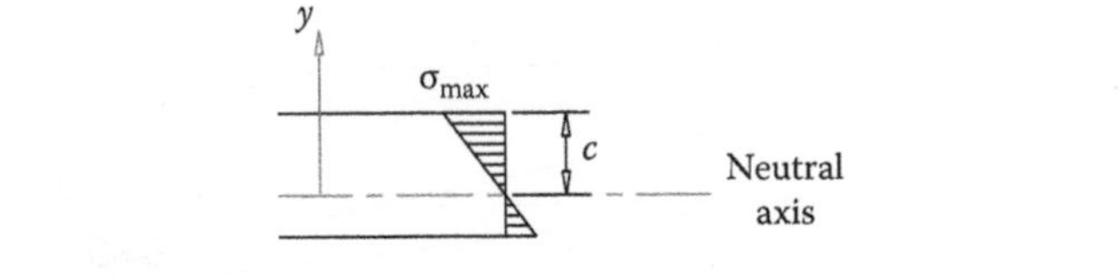

Figure 9.10 Bending.

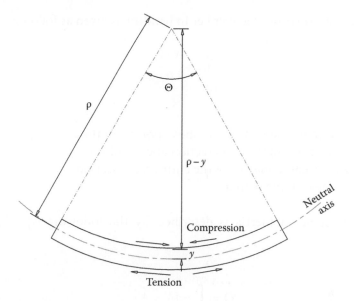

Figure 9.11 Beam in bending.

The stress and strain are negative values along the upper portion of the beam. This portion of the beam is in compression. The bottom of the beam is in tension, which produces positive stresses and strains.

From Hooke's law, the following can be determined:

$$\sigma_x = E\varepsilon_x = -\frac{y}{c}E\varepsilon_{max} = -\frac{y}{c}\sigma_{max} \qquad (9.41)$$

Through integration, it can be shown that these stresses are proportional to the applied moment:

$$\sigma_{max} = \frac{Mc}{I} \quad \text{and} \quad \sigma_x = \frac{My}{I} \qquad (9.42)$$

The I term is the area moment of inertia, which is equal to the section modulus S times the largest distance c from the neutral axis. Therefore,

$$\sigma_{max} = \frac{Mc}{I} = \frac{M}{S} \qquad (9.43)$$

The average shear stress of a member in bending is given as follows:

$$\tau_{av} = \frac{VQ}{It} \tag{9.44}$$

where
   Q is the first moment of inertia above point y in the cross section
   t is the thickness of the cross section above point y
   I is the moment of inertial of the entire cross section
   V is the shear in the section

The first moment of inertia is described by the following equation and Figure 9.12:

$$Q = \int_{y=y}^{y=c} y\,dA = A\bar{y} \tag{9.45}$$

In terms of the equation for average shear stress, the height of the cross section a is twice the largest distance c to the neutral axis. Additionally, the width of the member b is equal to t in the shear stress equation.

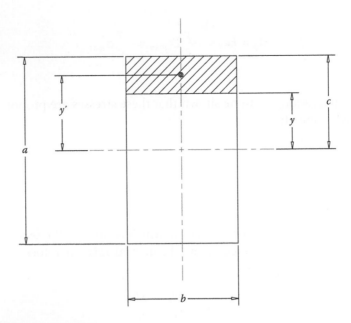

Figure 9.12 Cross section.

We define A' as the area of the hatched region, which has a centroid y' from the neutral axis:

$$Q = A'y'$$

$$A' = b(c - y)$$

$$y' = \frac{(c + y)}{2} \tag{9.46}$$

$$Q = \frac{1}{2}b(c + y)(c - y)$$

The area moment of inertia for a rectangular cross section can be modified to reflect these parameters:

$$I = \frac{1}{12}ba^3 = \frac{1}{12}b(2c)^3 = \frac{2}{3}bc^3 \tag{9.47}$$

Simplifying the shear stress equation, and noting that the total area A is the product of a times b, or two times the product of b times c, gives the following:

$$\tau_{xy} = \frac{VQ}{It} = \frac{3V}{4}\frac{(c^2 - y^2)}{1/2Ac^2} = \frac{3}{2}\frac{V}{A}\left(1 - \frac{y_2}{c^2}\right) \tag{9.48}$$

Since y is less than or equal to c, the maximum shear stress is given by the following:

$$\tau_{max} = \frac{3}{2}\frac{V}{A} \tag{9.49}$$

The maximum value of shear stress in a rectangular beam subject to bending is 50% larger than the uniform stress distribution ($\tau_{avg} = V/A$). Thus, a faulty assumption can easily result in an underestimation of maximum shear stresses.

Material Sizes of Humans

10

Introduction

In order to perform biomechanical computations, it is necessary to have some idea of the size of the structure that is to be analyzed. For example, if an engineer is asked to perform a calculation on the load on a beam, the material properties of the beam need to be known. These properties can be calculated, or more generally, are looked up in tables of the properties of the material, which is true because most structures have already been analyzed for their properties relative to their strength. Beams come in all shapes and sizes and are designed to carry a variety of loads. If a standard beam is not available, the engineer can design a particular beam to fit the application. In most instances, there is no need to reinvent the wheel by designing a particular beam, but rather applying a beam that will carry the specified loads, which may require utilizing a standard beam somewhat larger than necessary. To determine some basic dimensions for humans, we digress in order to study some characteristics of nature.

Living things on earth exhibit a great deal of symmetry. Actually, symmetry seems to permeate the known universe although we have only seen a very limited portion of the cosmos. Symmetry in nature is one of the fundamental principles in the search for a grand unified theory of the laws of physics. This concept is prevalent because we see symmetry not only in physical equations, in mathematics, and in biology but also in weather, in the planets, and in humans. With respect to human structure, we witness a great deal of symmetry from left and right, to two arms, two eyes, two legs, two lungs, and a host of other features. Artists, philosophers, and scientists dating back to the Greeks have noted this symmetry. This symmetry is not restricted to the number two. There are other forms of symmetry. A particular symmetry that appears in nature is represented by a particular number known as depicting beauty and proportion. This symmetrical proportionality number was utilized in architecture in ancient times and represented beauty in sculpture and is sometimes referred to as a golden ratio. This golden ratio of the Greeks (φ) can be closely approximated by the Fibonacci sequence, given as

$$0, 1, 1, 2, 3, 5, 8, 13, 21, 34, \ldots \tag{10.1}$$

This sequence is very simple where the following number is obtained by adding the previous two numbers. Thus, zero (0) plus one (1) equals one (1). One (1) plus one (1) equals two (2). One (1) plus two (2) equals three (3). Two (2) plus three (3) equals five (5), and so on. Mathematically, this sequence can be expressed as

$$x_n = x_{n-1} + x_{n-2} \qquad (10.2)$$

If you consider two successive Fibonacci numbers, the ratio is very close to the golden ratio of $\varphi = 1.618034\ldots$ The larger the pair of numbers used, the closer the approximation. The Fibonacci numbers may also be calculated from the golden ratio according to the following equation:

$$x_n = \frac{\varphi - (1 - \varphi)^n}{\sqrt{5}} \qquad (10.3)$$

The Fibonacci sequence can also be seen to create a spiral, as shown in Figure 10.1. This spiral is representative of a variety of structures found in nature. Some of these structures include sea shells, galaxies, weather, and human proportions.

The computations on loading effects on humans are somewhat different for a variety of reasons. First, humans come in all shapes and sizes within a range of limits. Although some humans may weigh as little as 4.7 lb at the age of 17 as was the case with Lucia Zarate. Lucia suffered from a severe case of primordial dwarfism and was from Mexico. At the other end of the spectrum, Manuel Uribe of Mexico weighed as much as 1230 lb. In terms of stature, the smallest person was also Lucia Zarate at 20 in. in height. It appears that Mexico has a propensity for extremes in terms of body dimensions. She lived between 1864 and 1890. The tallest man in recorded history was Robert Wadlow at a height of 8 ft 11.1 in. He lived from 1918 to 1940 in the United States.

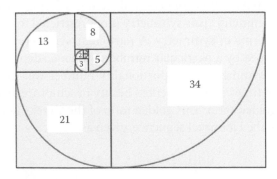

Figure 10.1 Fibonacci spiral.

Weights and Heights

Besides these extremes and others throughout recorded history, most people in the world fall within reasonable categories. In the United States, men and women, both white and black, are within the range of values as detailed in Tables 10.1 through 10.8. The tables give the range of heights and weights for the 95th, 50th, and 5th percentile. Tables 10.1 through 10.4 show weights while Tables 10.5 through 10.8 show heights.

Table 10.1   White Men's Weight by Age (lb)

Age	90th%	50th%	5th%
15	220	143	110
20	232	160	125
25	240	167	132
30	247	173	136
35	252	177	138
40	256	180	140
45	257	182	141
50	256	183	141
55	252	184	140
60	248	182	139
65	243	181	138
70	232	180	137
75	228	173	133
80	221	169	131

Table 10.2   Black Men's Weight by Age (lb)

Age	90th%	50th%	5th%
15	237	137	110
20	260	160	122
25	265	172	130
30	268	175	132
35	270	177	133
40	268	180	133
45	265	180	132
50	263	180	131
55	259	178	130
60	255	177	128
65	250	175	126
70	245	172	122
75	238	167	121
80	233	162	116

Table 10.3 White Women's Weight by Age (lb)

Age	90th%	50th%	5th%
15	192	120	97
20	210	130	103
25	220	132	106
30	227	139	107
35	231	142	108
40	233	149	110
45	234	151	111
50	230	152	112
55	228	153	114
60	228	153	113
65	224	151	111
70	220	148	110
75	214	143	108
80	210	138	105

Table 10.4 Black Women's Weight by Age (lb)

Age	90th%	50th%	5th%
15	221	132	96
20	235	142	108
25	242	150	110
30	253	158	111
35	258	164	112
40	263	170	113
45	264	172	117
50	264	174	118
55	262	174	118
60	258	172	118
65	251	170	114
70	246	165	111
75	242	158	110
80	234	150	107

Now, we know that some very tall people are quite thin while other short people are very heavy. A lot of the difference is due to excess adipose tissue (fat), but some of the variability is due to the supporting structure of the individual. Normally, the larger the individual, the more robust the skeleton is, because nature has designed individuals so that the supporting structure will accommodate the overall mass of the individual. The effects of excess mass on smaller frames is well understood and leads to a variety of ailments such as problems with the spine, the hips, and the knees to name a few. Again, it is not the purview of the forensic biomechanical engineer to diagnose

Table 10.5   White Men's Height by Age (in)

Age	90th%	50th%	5th%
15	73	69	63.5
20	74.5	69.5	65.5
25	75	69.6	66
30	75	69.8	66
35	74.5	69.8	66
40	74.5	69.8	65.5
45	74.2	69.8	65.5
50	74	69.5	65.3
55	73.5	69	65.2
60	73.2	69	65
65	72.8	68.5	64.8
70	72.5	68	64.5

Table 10.6   Black Men's Height by Age (in)

Age	90th%	50th%	5th%
15	73	68	63
20	74.3	69	64.8
25	75	69.3	65.2
30	75	69.7	65.2
35	74.5	69.9	65.2
40	74.5	70	65.2
45	74.2	70	65.2
50	74	69.5	65
55	73.5	69	64.7
60	73.2	68.7	64.2
65	72.8	68	63.8
70	72.5	67.8	63.2

the conditions, but it is within the realm of calculations and assessment to address the effects on the human structure of excess weight and abnormal dimensions.

Biomechanics can address the effect of the work environment on the individual. As we have pointed out, maladies such as tennis elbow and torn ligaments of the knees are often a result of the activities that we undertake. Plasterers and drywall hangers generally suffer from shoulder ailments because they constantly perform work overhead. Such work places excessive stresses on the shoulder joints leading to well-diagnosed conditions, including rotator cuff injuries and labral tears. Another example of work-related injuries occurs to flooring and carpeting installers. These folks often suffer from knee ailments resulting from their body weight impacting on those joints. Humans were simply not designed to use those structures in

Table 10.7   White Women's Height by Age (in)

Age	90th%	50th%	5th%
15	68.5	64	60.5
20	68.8	64.4	60.5
25	68.8	64.5	60.5
30	68.8	64.5	60.4
35	68.8	64.1	60.3
40	68.6	64	60.2
45	68.4	64	60.1
50	68	63.8	60
55	67.8	63.6	60
60	67.5	63.4	59.8
65	67.2	63	59.3
70	67	62.8	59

Table 10.8   Black Women's Height by Age (in)

Age	90th%	50th%	5th%
15	68	63.8	60
20	68.8	64.5	60.5
25	68.8	64.5	60.5
30	68.8	64.5	60.5
35	68.7	64.3	60.5
40	68.6	64.1	60.3
45	68.5	64	60.2
50	68.3	63.9	60.1
55	68	63.8	60
60	67.8	63.6	59.5
65	67.8	63.4	59.3
70	67.5	63.1	59

that manner. The knees were not designed to walk on. Other ailments stem from repetitive motion. Typists and computer workers are subject to carpal tunnel syndrome due to the repetitive movement of the fingers. They may also suffer from neck and shoulder problems resulting from bad posture while performing their work. Athletes suffer a variety of maladies to virtually every portion of their bodies from concussions, to broken bones, to torn ligaments and tendons, and worn cartilage in their joints. Repeated excessive loading of any structure is detrimental to the affected area.

In this chapter, we have attempted to include typical sizes for various body structures and the variability that is present. Knowing weight and height of the individual is the most helpful data since it is not feasible for biomechanical practitioners to conduct studies such as MRIs to assess the size of the injured human component. In some instances, this data may be available from health professionals. Generally, we need to know length, width, area, and volume of the structure. With these dimensions in hand, we are able to calculate the dimensional properties of these structures. These dimensional properties are the centroid, the first and second moments of inertia, the polar moment of inertia, and the radius of gyration. These concepts are fully explained in Chapter 11. For example, if a femur breaks as a result of an incident, and we wish to make some calculation on the force required to produce the injury, the length and the cross-sectional area are sufficient dimensions to make the calculation. Of course, we also need to know the strength of the material from Chapter 8.

Body Segments

Various researchers have studied the human body in order to quantify dimensions. Dempster and Gaughran have analyzed the properties of body segments based on their size and weight. Barter estimated mass of body segments in the 1950s. Previous work was also conducted on these topics by Bernstein and Braune in the 1930s and as far back as the 1880s by other researchers. Bogin and Varela-Silva have attempted to establish correlations between body proportions and health. Ruff has studied the effects of loading and stresses on body structures with emphasis on the long bones. A study by Elert and his students found that the proportion of height to wingspan in humans was 1.023 and the proportion of forearm plus hand to forearm was 1.715, which is close to the established value of 1.618, the golden ratio of the Greeks. The error in height to wingspan is about 2.3% and the error in the arm/hand measurements is about 6%. Both these values are well within engineering accuracy and validate Leonardo's Vitruvian Man where the wingspan and height form a square (see Figure 10.2).

In recent years, a significant emphasis has been placed on the determination of body proportions and segments. This work has become necessary because of the need for empirical data for the analysis and determination of the modes of locomotion of humans, biomechanical issues, construction of accurate test dummies, vehicular crashes and human falls, and the need for prosthetics as a result of injuries suffered during conflicts. The survival rate of casualties of war has spurred a very significant need for this type of analysis. In some instances of criminal activity, the biomechanical engineer may be called upon to perform calculations to prove or disprove certain allegations. Body dimensions may be obtained by a variety of methods, which include computer scans, dismemberment of cadavers, and application of natural laws such as the golden ratio. The values included for body dimensions were obtained from the studies cited in this chapter and are to be considered averages with the

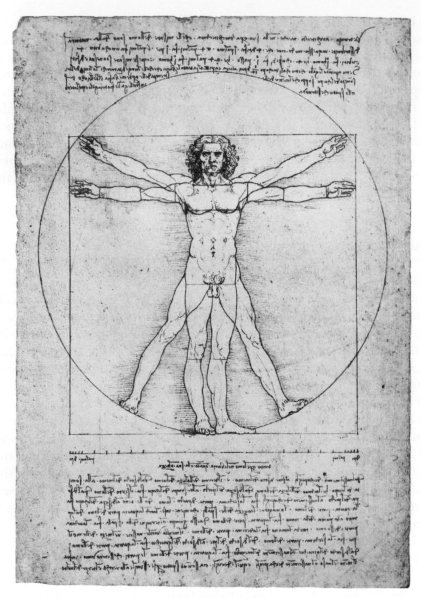

Figure 10.2 Leonardo's Vitruvian Man from 1492.

appropriate variability included. The values do not all add up to 100% because of the variability and the rounding by the authors. The error is insignificant. Table 10.9 shows body lengths in percentage of height from link to link. Links are essentially joint to joint. The link lengths across the shoulders are from the center of mass of the scapulas. The link lengths for the pelvis are from ileum to ileum. The tables include data from the Michigan study in 1963 by Krogman, Johnston, and Dempster. The Leipzig data is by Braune and Fischer.

Table 10.9 Body Lengths in % of Height from Link to Link

	Michigan Data	Leipzig Data
Vertical dimensions		
Foot length	3.7 ± 0.7	
Ankle to knee (shank)	25 ± 1.2	24.7 ± 0.9
Knee to hip	24 ± 1.4	24.2 ± 0.7
Hip to sternum	17.1 ± 2.0	
Sternum to shoulder	15.7 ± 1.5	
Shoulder to head CM	8.1 ± 1.3	
Shoulder to elbow	17.4 ± 1.0	17.9 ± 0.7
Elbow to wrist	15.7 ± 0.5	16.7 ± 0.8
Wrist to finger end	10.5 ± 0.6	
Horizontal dimensions		
Width of head	9.3 ± 0.6	
Shoulder to shoulder	18.3 ± 1.0	
Hip to hip	9.8 ± 1.2	

Table 10.10 Distance Between Proximal End and Segment cm

	Percent of Link	
	Michigan Data	Leipzig Data
Thigh	42.8 ± 2.0	43.7 ± 2.5
Shank	42.6 ± 2.3	42.1 ± 1.0
Arm	43.7 ± 3.0	45.9 ± 2.4
Forearm	42.9 ± 2.1	42.1 ± 1.1
Thorax	66.1 ± 6.4	
Lumbosacral segment	54.9 ± 7.5	

Table 10.10 shows the distance from the proximal end to the center of mass (CM) for the respective segment. Recall that proximal refers to the closest point of attachment to the body.

Table 10.11 represents the mean weights as a percentage of the total weights of male cadavers while Table 10.12 represents the mean volumes of these cadavers. The mass in percentage of the total weight is given in Table 10.13.

Some Mechanical Predictions

In biomechanical calculations, we attempt to predict the stresses and strains that produce injury to humans. The parameters that are important include the cross-sectional area of the particular material C_a, the mass moment of inertia I, which is the torque required for a desired angular acceleration

Table 10.11   Mean Weights of Male Cadavers

	Weights as % of Total Weight	
	Michigan Data	Leipzig Data
Head and trunk	56.3 ± 2.5	52.2 ± 3.0
Head and neck	7.9 ± 0.9	
Arm	2.6 ± 0.3	3.2 ± 0.3
Forearm	1.5 ± 0.2	2.1 ± 0.25
Hand	0.61 ± 0.06	0.8 ± 0.05
Thigh	10.0 ± 1.2	10.9 ± 0.77
Shank	4.6 ± 0.53	4.7 ± 0.35
Foot	1.4 ± 0.14	1.8 ± 0.2

Table 10.12   Mean Volume of Cadaver Segments

	Ratio to Body Volume	
Body Section	Volume (cm³)	Ratio
Thigh	5,865 ± 969	9.588 ± 1.142
Shank	2,643 ± 654	4.221 ± 0.502
Arm	1,533 ± 332	2.439 ± 0.294
Forearm	857 ± 180	1.531 ± 0.167
Thorax	7,752 ± 1978	12.109 ± 1.370
Hand	342.7 ± 73.8	0.612 ± 0.057
Foot	833 ± 174	1.294 ± 0.143
Abdominopelvic mass	15,854 ± 2366	26.274 ± 3.361
Shoulders	3,172 ± 919	5.063 ± 0.563
Head and neck	4,508 ± 818	7.178 ± 0.709
Head and trunk	33,970 ± 4570	45.545 ± 1.377
Total volume	61,190 ± 8137	

Table 10.13   Mass of Body Components

	Percent of Total Weight		
Body Section	Skin and Fascia	Muscle	Bone
Thigh	29.0 ± 3.31	59.6 ± 2.81	11.5 ± 1.78
Shank	22.2 ± 2.98	45.6 ± 2.85	32.3 ± 2.56
Arm	26.0 ± 3.85	56.8 ± 4.74	18.0 ± 1.61
Forearm	18.7 ± 4.25	53.2 ± 4.90	28.2 ± 3.99
Foot	29.2 ± 4.30	20.3 ± 4.25	50.7 ± 7.35
Hand	28.4 ± 3.04	27.6 ± 2.98	44.0 ± 3.27
Left shoulder	36.5 ± 3.34	53.9 ± 4.09	9.6 ± 3.76

about a particular axis of the body, and the polar moment of inertia J, which is the property that allows for the determination of the deflection of the particular body. As an example, the average cross-sectional area of a human disc is approximately 660 mm². This value may vary somewhat between the ranges of 600 and 720 mm² for a total variation of 10% in either direction. The cross-sectional area of a human femur on average is 3.125 cm². Varying this area by 10% in either direction yields a value between 2.81 and 3.44 cm². The moment of inertia allows for the determination of compressive and tensile stresses, while the polar moment of inertia is used in determining bending and torsional stresses. Empirical tests for human long bones by Ruff showed that there is a relationship between the moment of inertia I, the perpendicular distance to the specified axis r and the polar moment of Inertia J. Additionally, the axial stress σ_a and the bending stress σ_b are given by the following:

$$\sigma_a = \frac{F_a}{C_a} \ (\text{lb/in.}^2 \text{ or kg/m}^2) \tag{10.4}$$

$$\sigma_b = \frac{r M_b}{I} \ (\text{lb/in.}^2 \text{ or kg/m}^2) \tag{10.5}$$

where
  F_a is the axial force (lb or kg)
  M_b is the bending moment (ft lb or kg m)

For any two perpendicular axes, the sum of the moments of inertia is equal to the polar moment of inertia. These empirical tests determined that

$$I^{0.73} = \frac{I}{r} \approx J^{0.73} \tag{10.6}$$

In those studies, the applied force for both axial and bending was determined to be proportional to body weight and it was predicted that the cross-sectional area will scale with body mass and that the polar moment $J^{0.73}$ will scale with the product of body mass and bone length. The conclusion of these studies determined that scaling factors for upper and lower limb bones are similar and are related according to the relationship exemplified in Figure 10.3.

In Figure 10.3, the vertical scale is the logarithm of the average strength ($J^{0.73}$) and the horizontal scale is the logarithm of the product of the body mass multiplied by the bone length. These studies indicate that the mechanical scaling factors for upper and lower limb bones are similar and that the

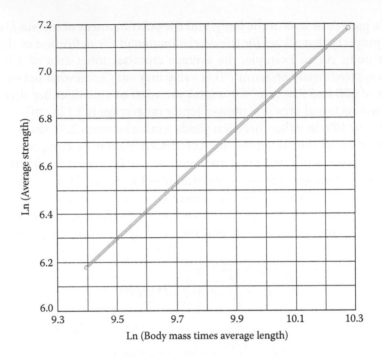

Figure 10.3 Bone scaling.

same scaling factors can be used for both. The product of body mass and bone length represents an adequate representation of bending and torsional moments that may be applied to the long bones of humans. This type of analysis can be used to approximate bone length for injury causation when body mass of the individual is known but stature is not known.

Ligaments, Tendons, and Cartilage

Probably the most common tendon injury is to the rotator cuff in the shoulder girdle. McFarland et al. have classified rotator cuff tears in terms of body dimensions as outlined in the following. The rotator cuff tendon attaches to the humeral head. This tendon begins to wear out around the age of 30 as a result of wear and tear. By the age of 50, most people have at least some form of wear in the rotator cuff tendons. This malady affects to a greater degree people who overuse the shoulder through overhead movements. Tendon tears are classified by the size of the tear whether it is partial or complete. The rotator cuff tendon varies in size according to the size of the individual. As a rough estimate, the tendon is approximately as wide as three fingers of the individual. Thus, it varies from about 6 to 8 cm in width. A small tear would be less than 1 cm. A moderate tear would be 1 cm × 3 cm. A large to massive tear means that the full width of the tendon is torn and the tendon may be completely detached from the bone. Tendons fail around a strain of between 12% and 15% and a stress between 100 and 150 MPa.

According to an anatomic study of the supraspinatus insertion and footprint by Ruotolo, Fow, and Nottage, a correlation between the normal cuff thickness and its humeral head attachment was carried out to determine the amount of tendon loss. They determined that the mean anteroposterior dimension of the supraspinatus insertion was 25 mm. The mean superior to inferior tendon thickness at the rotator interval was 11.6 mm, 12.1 mm at mid-tendon, and 12 mm at the posterior edge. The distance from the articular cartilage margin to the bony tendon insertion was 1.5–1.9 mm with a mean of 1.7 mm. In their conclusion, they determined that tears showing exposed bone greater than 7 mm laterally to the articular margin approximated 50% of the tendon substance.

Another common tendon injury is to the Achilles tendon. People involved in sports are susceptible to this injury. Pang and Ying studied 40 healthy men and women of different ages to determine the properties of Achilles tendons. Table 10.14 summarizes their findings.

Cartilage is the lubricating surface between bones that move relative to each other. Cartilage is found in the hips, knees, elbows, shoulders, ankles, and the other joints of the hand, feet, fingers, and toes. The most common injury to cartilage occurs to the knee joint as a result of stresses placed on the rim of the cartilage. Often times, torn cartilage occurs because of an anomaly to the weight-bearing area of the medial and lateral femoral condyles. The condyles are the projections of the lower portion of the femur and are the weight-bearing surfaces. The medial condyle is larger than the lateral condyle because it supports a larger portion of the weight of the individual as the center of gravity is medial rather than lateral. Consequently, there are more injuries to the cartilage medially rather than laterally. Guettler et al.

Table 10.14 Achilles Tendon Dimensions

Characteristic	Thickness (mm)	Cross-Sectional Area (mm²)	Length (mm)
Age			
20–29	5.14 ± 0.57	57.8 ± 11.61	111.16 ± 29.39
30–39	4.86 ± 0.69	55.3 ± 13.96	119.37 ± 24.48
40–49	5.12 ± 0.62	59.5 ± 11.44	121.26 ± 21.60
Greater than 50	5.28 ± 0.64	70.5 ± 15.33	119.13 ± 18.73
Ankle dominance			
Nondominant	5.11 ± 0.65	58.33 ± 14.25	119.72 ± 22.87
Dominant	5.09 ± 0.63	63.23 ± 13.89	115.74 ± 24.71

determined that cartilage rim defects less than 8 mm did not demonstrate stress concentrations causing progression of the defect. For anomalies greater than 10 mm, peak pressures progressed the rim size of the defect. As the defects were enlarged by the application of 700 N of pressure, the defect size increased from 10 to 20 mm. The mean distance of the progression of the defect varied from 2.2 mm on the medial condyle and 3.2 mm on the lateral condyle.

Ligament injuries occur most often to the knee joint because of its weight-bearing function. These injuries are produced by activities that twist the knee, blows to the knee, hyperextending the knee, sudden stops while running, jumping, and shifting of the body weight. There are four ligaments in the knee. These are the anterior cruciate ligament (ACL) that connects the thigh bone (femur) to the shin bone (tibia), the posterior cruciate ligament (PCL) that also connects the femur to the tibia, the lateral collateral ligament (LCL) that connects the femur to the fibula, and the medial collateral ligament (MCL) that connects the femur to the medial side of the tibia. The most common injury occurs to the ACL. The PCL is approximately twice as strong and thick as the ACL.

Most knee problems arise as a result of the aging process and the associated stresses and strains of daily living. These conditions are referred to as arthritis. Arthritis is the wearing away of the cartilage in the knee or any other joint. This damage is compounded by inflammation to the joint. Recall that the meniscus is the cartilage at the end of the knee joint on the wear surfaces that lubricates and acts as a shock absorber for the joint. Knee problems may also arise as a result of impact or rotational insult to the affected areas. Most meniscus tears occur while the knee bears weight and the upper body twists as the foot stays in place. Seldom are torn cartilage problems associated with impacts without the knee being in a weight-bearing mode. Car accidents are not responsible for cartilage tears because the occupants are seated and, their knees not bearing weight. This fact is especially true for low impact collisions.

Stone et al. performed a study to correlate the size of the meniscus based on gender, height, and weight. They studied both the lateral and the medial menisci with respect to the width and the length. The length measurement was anterior posterior and the width was lateral medial. The total plateau width varied from approximately 6 cm for stature heights of 60 in. to approximately 9 cm for stature heights of 77 in. in a relatively linear fashion. For low body mass index (BMI) individuals, the total plateau width was 7.37 cm and for high BMI individuals the total plateau width was 7.84 cm. The lateral meniscus width varied from 2.84 to 2.98 cm. The lateral meniscus length varied from 3.48 to 3.70 cm. The medial meniscus width varied from 2.95 to 3.22 cm. The medial meniscus length varied from 4.38 to 4.51 cm. The greater values are for the higher BMI individuals.

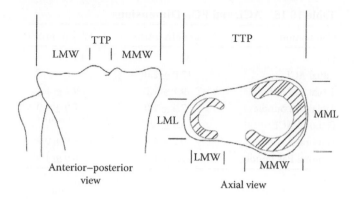

Figure 10.4 Meniscus.

Please refer to Figure 10.4. The cutoff for BMI measurements is according to the Center for Disease and Control and Prevention at a value of 25. Recall that BMI is defined as

$$BMI = 0.703 \times (weight \ (lb)/height \ (in.)) \tag{10.7}$$

Another study by Bloecker et al. indicates that meniscus surface area strongly scales with tibial plateau area for both sexes.

Injuries to the knee ligaments may be associated with sports or with collisions. The most common ailment is to the ACL, which controls rotation and forward movement of the tibia. Football, basketball, skiing, and hockey are common sports that injure this ligament. The PCL controls the backward movement of the tibia and is commonly injured with a direct impact or sudden blow from a car accident or a football tackle. The other ligaments of the knee may also be injured. The MCL gives stability to the inner knee while the LCL gives stability to the outer knee.

Dimensions for the ACL and the PCL are given in Table 10.15. The dimensions for the ACL are much more detailed than that for the PCL because it has been studied to a much greater degree because of its greater propensity for injury.

Kopf et al. found a large variation in ligament sizes and a very weak correlation between knee ligaments, height, and weight of the individuals tested based on 137 patients. Consequently, height and weight are not good predictors of ligament dimensions. These studies were undertaken as a result of knee injuries to the affected individuals. The ACL is comprised of two bundles, the posterolateral (PL) and the anteromedial (AM) bundles attaching the tibia to the femur. Either of these bundles may be injured.

Table 10.15   ACL and PCL Dimensions

Connection	Length (mm)	Width (mm²)
PCL	38	13
Tibial ACL	17.0 ± 2.0	
Tibial AM bundle	9.1 ± 1.2	9.2 ± 1.1
Tibial PM bundle	7.4 ± 1.0	7.0 ± 1.0
Femoral ACL	16.5 ± 2.0	
Femoral AM bundle	9.2 ± 1.2	8.9 ± 0.9
Femoral PM bundle	7.1 ± 1.1	6.9 ± 1.0

Bones

Various researchers have analyzed the relationship between bone length and bone cross-sectional area in humans and hominids. These studies have been carried out for the lower and the upper extremities and also the metacarpals and metatarsals. The greater emphasis has been on the leg and arm bones because of their significance with respect to injury. Schiessl and Willnecker analyzed lower leg measurements with computed tomography (CT). O'Neill and Ruff estimated human long bone cross-sectional properties. Kntulainen et al. analyzed cortical bone cross-sectional geometries. Marchi and Tarli as well as many others have studied the cross-sectional geometries of limb bones and their effect due to loading.

The cross-sectional area of the femur can be variable according to the stature of the individual and the weight it is designed to support. As a central value, it can be estimated that the cross-sectional area of the femur of humans is 3.6×10^{-4} m². The ultimate strength of bone tissue is typically between 120 and 170 N/m². It is well known that when bone is compressed, it is shortened. The difference in the length relative to the original strength is the strain. Under loading below a certain threshold of 800 microstrains, bone loss occurs. Between 800 and 1600 microstrains bone is preserved. Above 1600 microstrains, bone is added. Thus, high-impact training rather than endurance training adds bone and muscle mass. Bone and muscle cross-sectional areas have a high degree of correlation. Of course, men have a higher ratio of bone and muscle than women. Men of any age and for women older than 45 years of age 20 cm² of muscle produce 1 cm² of bone. For young women approximately 16 cm² of muscle produce 1 cm² of bone. Figure 10.5 shows this relationship. The length of the bone area was measured at 66% if the tibial length proximal to the ankle joint.

Cortical bone, sometimes called compact bone, is the main support structure for the human anatomy. Cortical bone, whose name is derived from cortex or outer area of most bones, comprises approximately 80% of the total weight of the human skeleton. The outermost layer of bones is covered

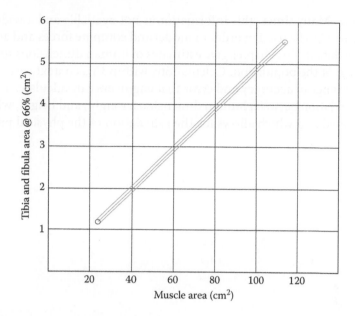

Figure 10.5 Muscle and bone mass.

with a membrane called the periosteum. The inner cavity of bone is called the medullary cavity and is filled with bone marrow. Measurements on cadaver tibias indicate that for a mean total area of the bone of approximately 383 mm² the cortical area comprises approximately 208 mm². Therefore, the ratio of total area to hard bone support area of bone is approximately 54%. A good mechanical model for long bones is a hollow elliptical cross section for analysis of the breaking strength of bones. Standard mechanical analysis techniques can then be used on the cross-sectional model to determine the breaking strength of these bones. In particular for the human femur and tibia, the cross-sectional properties derived utilizing the hollow elliptical model are very well correlated with the true properties with errors between 3% and 8%.

Summary

Estimation of human proportions and dimensions is not an exact science when performing biomechanical calculations. There are methods of analysis such as CT, x-rays, peripheral quantitative computed tomography (pQCT), and others to more accurately predict the dimensions of a particular body part. However, the biomechanical engineer normally does not have access to the patient or to these diagnostic techniques. For biomechanical calculations on the possibility for injury, at best, the weight and stature of the individual is available. Many times, this information is not available and a wide range of values need to be used in order to model and compute forces and accelerations. The data in this chapter give estimates of human dimensions for a very large range of the population. Calculations within this context certainly fall under the range of accuracy and error that engineers can calculate. There is a direct correlation between human dimensions, weight, and height, which we have outlined and which allows for the calculation of the physical processes of injury.

Statics and Dynamics 11

Newton's Laws

The motion of a particle was first correctly described by Sir Isaac Newton in terms of three fundamental laws. These laws have been verified through measurements and tests conducted in the ensuing centuries. These laws are as follows:

> *Newton's first law*: A particle remains at rest or continues to move in a straight line and uniform velocity if there is no unbalanced force acting upon it.
>
> *Newton's second law*: The acceleration of a particle is proportional to the resultant force acting on it and it occurs in the direction of this force.
>
> *Newton's third law*: The forces of action and reaction between interacting bodies are equal in magnitude, opposite in direction, and collinear.

In a biomechanical analysis of the motion of an occupant's head in a rear-end vehicular collision, these laws are useful in determining important parameters (i.e., force, acceleration, and propensity for injury). According to the first law, an occupant's head will continue to move (rotate and translate) in response to a collision until acted on by a force. This force may be associated with the back of the head striking the headrest in a rear-end collision, or the internal resistances (forces) to motion that occur within the neck structure.

The second law simply states that force equals mass times acceleration, or

$$F = ma. \tag{11.1}$$

The bolded notations indicate that the force "*F*" and acceleration "*a*" are vector quantities. Thus, they have both magnitude and direction. In the example earlier, the force acting on the occupant's head as it strikes the headrest is equal to the mass of the occupant's head times its acceleration. The acceleration in this case is dependent on the speed change associated with the collision and the deceleration distance during the impact with the headrest.

This deceleration distance depends on the compression of the cushioned headrest, as well as the internal compression of bodily structures (i.e., skin and muscles).

The third law implies that the forces in a collision are equal and opposite in direction. Thus, the force acting on the occupant's head is the same as the force acting on the headrest. This information may be useful in determining failure loads for the headrest or injury potential for the occupant.

The acceleration of a vehicle or occupant in a collision is often listed in terms of g's, or gravitational units. One "g" is equal to the acceleration due to gravity at sea level.

$$g = 32.2 \text{ ft/s}^2 = 9.81 \text{ m/s}^2.$$

At one "g," the force "F" acting on a body is equal to its weight "W"

$$F = ma = mg = W. \tag{11.2}$$

In SI units, the mass is in kilograms (kg) and the force is in Newtons (N), while English units utilize slugs (lb s^2/ft) for the mass quantity and pounds (lb) for the force.

Force Systems and Components

As noted earlier, forces are vector quantities in that they possess a magnitude and direction. Forces can be broken down into components according to the chosen coordinate system. In a rectilinear or Cartesian coordinate system, the components of a force vector occur along the 3D axes (Figure 11.1).

Summing these forces in vector notation is shown as follows:

$$\boldsymbol{F} = F_x \boldsymbol{i} + F_y \boldsymbol{j} + F_z \boldsymbol{k} \tag{11.3}$$

Here, $\boldsymbol{F}$ is the vector sum of the force components and the unit vectors $(\boldsymbol{i}, \boldsymbol{j}, \boldsymbol{k})$ in the x-, y-, and z-axes. The magnitude of the force is given by the following:

$$|F| = \sqrt{F_x^2 + F_y^2 + F_z^2} \tag{11.4}$$

In an analysis of long bones, it may be more useful to use cylindrical coordinates (Figure 11.2).

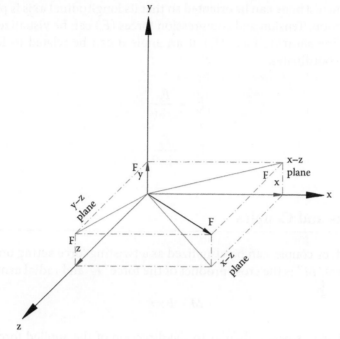

Figure 11.1 Forces in Cartesian coordinates.

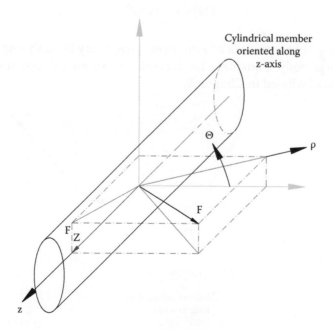

Figure 11.2 Forces in Cylindrical coordinates.

Forensic Biomechanics and Human Injury

As shown, a bone can be oriented so that its longitudinal axis is placed in the z-direction. Tension and compression forces (F_z) can be visualized along this axis. The shearing force (F_ρ) at an angle θ can be related to forces in Cartesian coordinates:

$$F_\rho = \frac{F_x}{\cos\theta}$$

$$(11.5)$$

$$F_\rho = \frac{F_y}{\sin\theta}$$

Moments and Couples

A moment, or couple, can be visualized as a twisting force acting on a body. The moment "M" is the cross product of the force "F" and radial arm "r":

$$M = F \times r \qquad (11.6)$$

The radial arm is perpendicular to the direction of the applied force. Thus, the magnitude of the moment, as described in Figure 11.3, is given by the following:

$$|M| = F \cdot r \cdot \cos\theta \qquad (11.7)$$

Determining the magnitude of a moment is necessary in analyzing torsional stresses on a body. Equations for torsion on cylindrical and noncircular members are included in Chapter 9.

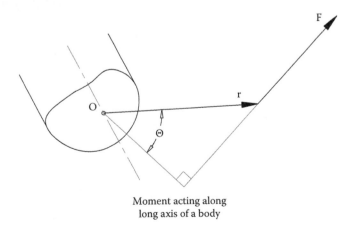

Moment acting along
long axis of a body

Figure 11.3 Moment arm.

Equilibrium

The principle of equilibrium is an essential tool in computing forces and moments acting on a biological structure. It follows from Newton's third law, which states that every action has an equal and opposite reaction, that forces in a specific direction will add to zero. For a Cartesian coordinate system, this statement indicates the following:

$$\sum F_x = 0$$

$$\sum F_y = 0 \qquad (11.8)$$

$$\sum F_z = 0$$

With the exception of the simplest problems, summing forces in each direction may not provide enough equations to determine all of the unknowns. In order for a structure to be statically determinate, the number of equations must equal the number of unknowns. Computing moments about a point in the structure provides additional equations for analysis.

Varignon's Theorem states that the moment of a force about any point is equal to the sum of moments of the force components about the same point. Thus, the sum of moments about any point zero on a structure is equal to zero:

$$\sum M_o = 0 \qquad (11.9)$$

These basic principles, along with the construction of an accurate free-body diagram, form the basis of any biomechanical analysis.

Free-Body Diagrams

Figure 11.4 displays forces acting through a human foot in a standing (prone) position and during a calf-raise maneuver. For the prone position, the normal force through the toe is assumed to be zero. This assumption is both reasonable and necessary for the problem to be statically determinate. In the calf-raised position, the load is taken off of the heel, and normal loads are present at the toe. For each of these situations, a free-body diagram can be developed to determine the effect of movement on the normal loads acting on the base of the foot.

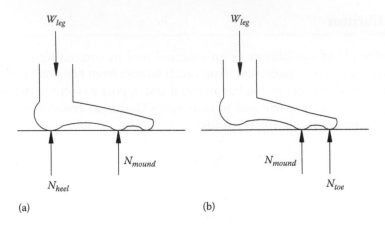

Figure 11.4 Forces acting through a foot: (a) prone position and (b) raised position.

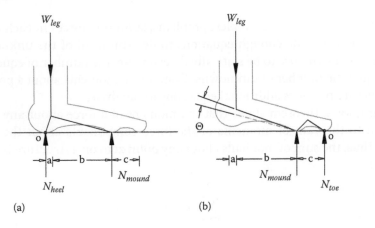

Figure 11.5 Free-body diagrams: (a) prone position and (b) raised position.

Figure 11.5 details the free-body diagrams for each case. The horizontal dimension "a" is the distance between the center of the heel and the downward force acting through the leg. The dimension "b" is the distance between this load and the foot mound. The distance between the mound and toe is referred to as "c."

For the prone position, moments are taken about the heel, which permits a determination of the force acting on the foot mound in terms of the downward force and foot dimensions:

$$\sum M_O = 0 : W_{leg} \times a - N_{mound} \times (a+b) = 0$$

$$N_{mound} = W_{leg}\left(\frac{a}{a+b}\right)$$

(11.10)

To determine the normal load acting on the heel, forces are summed in the vertical direction:

$$\sum F_y = 0 : N_{heel} + N_{mound} - W_{leg} = 0$$

$$N_{heel} = W_{leg}\left(1 - \frac{a}{a+b}\right)$$

(11.11)

Here, we have normal forces acting in the upward (positive-y) direction. We can see that these normal loads are a fraction of the total weight acting on the foot.

The raised position presents a shift in loading to the front of the foot. Assuming a small lift angle "θ," the horizontal position "b" remains essentially unchanged. For greater extension of the ankle, this distance becomes "$b \times \cos\theta$," and thus the lift angle must be taken into account:

$$\sum M_O = 0 : -W_{leg} \times (b+c) + N_{mound} \times c = 0$$

$$N_{mound} = W_{leg}\left(\frac{b+c}{c}\right)$$

(11.12)

Summing forces about the vertical direction permits determination of loading through the toe:

$$\sum F_y = 0 : N_{mound} + N_{toe} - W_{leg} = 0$$

$$N_{toe} = W_{leg}\left(1 - \frac{b+c}{c}\right)$$

(11.13)

Here, we see that "N_{mound}" acts in the upward y-direction, while "N_{toe}" is directed downward. Additionally, it is evident that the normal load acting through the mound of the foot (N_{mound}) is greater than the total load (W_{leg}) acting through the foot.

The types of connections and interactions are important to correctly identify in a free-body diagram. These connections may involve hinges, in which normal loads are developed. However, moments are not produced in hinge-type connections, as rotation is freely permitted. Figure 11.6 through Figure 11.9 describes various connection types that may be encountered in a biological structure.

Axial force systems shown in Figure 11.6 include tension members, such as tendons and ligaments. Tendons provide a pulling force, which permits

Forensic Biomechanics and Human Injury

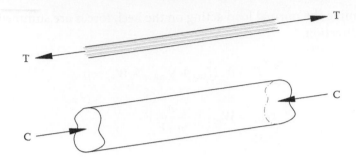

Figure 11.6 Axial force systems.

movement relative to a joint. The loads on these members act along the longitudinal axis. Similarly, compression members, such as rigid bones, develop stresses
along the long axis, albeit in an opposite direction than tension members.

Contacting surfaces may be examined internally, such as at joints, and
externally, such as the interaction between a foot and walking surface. The
compression of cartilage or the slip resistance of a walking surface may be
examined through such a diagram. Figure 11.7 shows that additional loads
are developed at the interaction between rough surfaces. The force "f" is
related to the normal load "N" by a ratio "μ," which is known as the coefficient of friction.

$$f = \mu N \tag{11.14}$$

As noted earlier, pinned connections involve free rotation. In contrast,
moments are developed at fixed connections. Within the human body, most
connection points between bones are essentially hinged, although tendons
and ligaments limit the range of motion. Fixed connections are not normally considered unless an element of the body is externally restrained
(Figure 11.8).

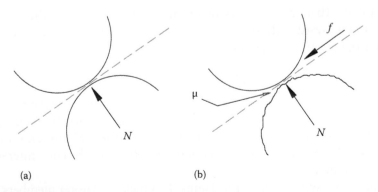

(a)   (b)

Figure 11.7 Contacting surfaces: (a) smooth surfaces and (b) rough surfaces.

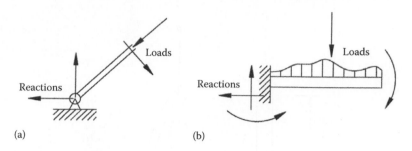

Figure 11.8 Contacting surfaces: (a) pinned and (b) fixed connections.

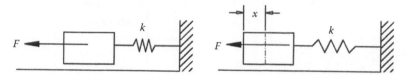

Figure 11.9 Spring force.

The spring force may involve stretching of a muscle or tendon below its elastic limit. As noted in Chapter 9, the elastic response of materials is linear below the yield point. This principle allows the determination of force "F" in terms of a spring constant "k" and displacement distance "x" (Figure 11.9):

$$F = kx \qquad (11.15)$$

Frames and Force Systems

Biological structures, especially the skeletal system, may be analyzed as frames. A frame is a rigid structure that assists in carrying a load. Typical frames are composed of beams, columns, tension bars, and trusses. In a skeletal system, truss structures are not encountered, and thus may be ignored.

As outlined in Chapter 9, compression and tension members develop stresses proportional to the cross-sectional area:

$$\sigma_c = \frac{C}{A}; \quad \sigma_t \frac{T}{A} \qquad (11.16)$$

It may be useful to determine the critical buckling load of a column. A compressive member is considered to be stable as long as the load does not exceed this critical value. First, we consider a column with pinned connections at each end. As noted earlier, a pinned connection is one that restrains against translational movement, but permits rotation (Figure 11.10):

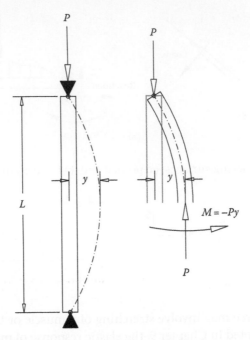

Figure 11.10 Buckling load on pinned–pinned column.

Recall from Chapter 9 the figure for the curvature of a beam under loading. The inverse of the beam curvature is equal to the second derivative of the beam's deflection:

$$\frac{1}{\rho} = \frac{M}{EI} = \frac{d^2y}{dx^2} \tag{11.17}$$

Here
"M" is the moment
"E" is the modulus of elasticity
"I" is the moment of inertia of the beam's cross section

For a compressive member under a load of "P," the moment can be rewritten as the product of this load times the maximum deflection "y":

$$\frac{d^2y}{dx^2} = \frac{M}{EI} = -\frac{Py}{EI} \tag{11.18}$$

The negative term is chosen for the purpose of simplicity, as rearranging this equation yields a linear, second-order, homogeneous differential equation:

$$\frac{d^2y}{dx^2} + \frac{P}{EI}y = 0 \tag{11.19}$$

The general solution to this equation is given as follows:

$$y(x) = A_1 \sin\left(\left(\sqrt{\frac{P}{EI}}\right)x\right) + A_2 \cos\left(\left(\sqrt{\frac{P}{EI}}\right)x\right)$$ (11.20)

The boundary conditions dictate that for a member of length "L," $y(x=0)$ and $y(x=L)$ should be equal to zero. Thus, the critical buckling load "P_{cr}" will be as follows:

$$P_{cr} = \frac{\pi^2 EI}{L^2}$$ (11.21)

The buckling load is dependent upon the manner of connections at each end. These connection types alter the boundary conditions, which affects the proposed solution. For a cantilevered column with a fixed connection at one end, consider Figure 11.11.

As shown, the effective length "L_e" is twice the actual length "L" of the member. This relationship is controlled by the effective length factor "K" for columns:

$$L_e = KL$$ (11.22)

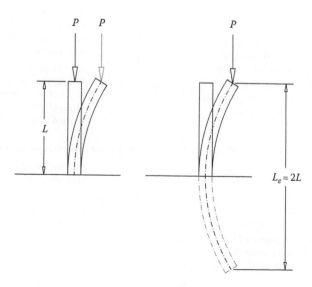

Figure 11.11 Buckling load on cantilevered column fixed at its base.

The theoretical values for "K" are given in the following table. Additionally, design values for "K" utilized by structural engineers are listed.

Type of Connections	Theoretical "K" Value	Design "K" Value
Fixed–fixed	0.50	0.65
Pinned–fixed	0.70	0.80
Fixed–free translation	1.00	1.20
Pinned–pinned	1.00	1.00
Fixed–free translation and rotation	2.00	2.10

In light of this relationship, we may update the equation for buckling load as such:

$$P_{cr} = \frac{\pi^2 EI}{L_e^2} = \frac{\pi^2 EI}{(KL)^2} \tag{11.23}$$

Beams are structural members, normally oriented in a horizontal direction, which support gravity loads. These loads are applied in a transverse direction to the beam. Beams typically span between supports and may include cantilevers beyond the supported ends. The supports may be considered fixed, pinned, or restrained in only one direction. The development of equations for moments and shearing loads on beams is beyond the scope of this book. However, Figure 11.12 provides basic equations for commonly encountered beams in the following:

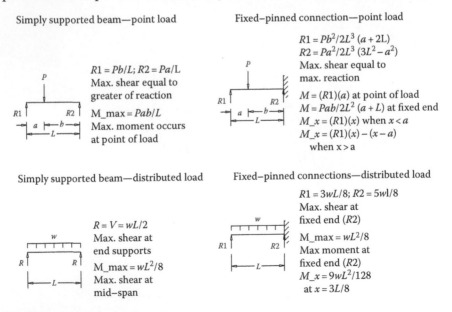

Simply supported beam—point load

$R1 = Pb/L; R2 = Pa/L$
Max. shear equal to greater of reaction
$M\_max = Pab/L$
Max. moment occurs at point of load

Fixed–pinned connection—point load

$R1 = Pb^2/2L^3 (a + 2L)$
$R2 = Pa^2/2L^3 (3L^2 - a^2)$
Max. shear equal to max. reaction
$M = (R1)(a)$ at point of load
$M = Pab/2L^2 (a + L)$ at fixed end
$M\_x = (R1)(x)$ when $x < a$
$M\_x = (R1)(x) - (x - a)$ when $x > a$

Simply supported beam—distributed load

$R = V = wL/2$
Max. shear at end supports
$M\_max = wL^2/8$
Max. shear at mid-span

Fixed–pinned connections—distributed load

$R1 = 3wL/8; R2 = 5wl/8$
Max. shear at fixed end $(R2)$
$M\_max = wL^2/8$
Max moment at fixed end $(R2)$
$M\_x = 9wL^2/128$
at $x = 3L/8$

Figure 11.12 Beam formulas.

In a biological structure, the purpose of a beam analysis may not be obvious. However, such an analysis may be useful when examining the effects of transverse loading on an individual bone. Similarly, the reaction and couple applied to the fixed end of a beam may assist in determining failure modes of a ligament or tendon.

A series of rigid, connected members formed to transfer forces and support a load is referred to as a structure. Structures include trusses, frames, and machines. Simple, plane trusses are composed of two-force members that develop stresses in tension or compression. Joints in trusses are pinned connection, wherein moments are not developed at each joint. Frames and machines are structures in which one or more components are multiforce members. Multiforce members are acted on by at least three forces, or by two forces and at least one moment. Frames support static loads. Machines typically include moving parts and are designed to transfer input forces and moments to output forces and moments.

As previously stated, truss structures are not normally encountered in a biomechanical analysis. Thus, we concentrate on frames and machines. As with trusses, frames and machines are analyzed by constructing free-body diagrams of each part, which is similar to methods of joints and sections used to determine the magnitude of two-force components within a truss. Newton's third law must be observed in such an analysis, as the action on one component must be equal to the reaction of a connecting component. However, the structures must first be analyzed by considering all forces external to the structure by visualizing the frame or machine as a single, rigid body.

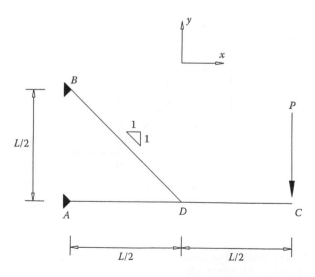

Figure 11.13 Rigid frame.

Consider Figure 11.13, which represents a rigid frame loaded vertically at its end.

This frame is considered rigid and noncollapsible. The frame may represent the application of gravity loads at the end of an arm. The segment *BD* provides tension support to the frame, and resists moments produced by the load across the cantilevered arm. In order to analyze this structure, equilibrium conditions are set across the entire frame. Then, the individual members are analyzed, with the tension load across segment *BD* of particular concern. This member may represent static tension loads on a tendon. Figure 11.14 details the free-body diagrams necessary for analyzing this structure:

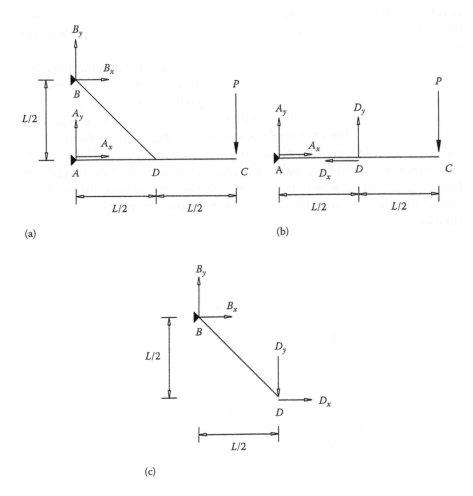

Figure 11.14 Free-body diagrams for rigid frame: (a) external reactions, (b) member *AC*, and (c) Member *BD*.

For the external reactions, summing forces in each direction yields the following relationships:

$$\sum F_x = 0 : A_x + B_x = 0 \rightarrow A_x = -B_x$$

$$\sum F_y = 0 : A_y + B_y - P = 0 \rightarrow A_x + B_x = P$$

(11.24)

Summing moments about point A and B provides the x-direction reactions at these points:

$$\sum M_A = 0 : PL + B_x \frac{L}{2} = 0 \rightarrow B_x = -2P$$

$$\sum M_B = 0 : PL - A_x \frac{L}{2} = 0 \rightarrow A_x = 2P = -B_x$$

(11.25)

As shown, the relationship between A_x and B_x is confirmed. Note that the direction of B_x should be to the left, which is in contrast to the initial free-body diagram.

For the member AC, the direction of reactions at point D is chosen to represent tension loading. For frame and truss analyses, loads pointing away from a joint indicate that the member is in tension. Moreover, it is intuitive that the vertical reaction at point D (D_y) will be directed upward to counter-act the moment produced by the gravity load P:

$$\sum F_x = 0 : A_x - D_x = 0 \rightarrow A_x = D_x = 2P$$

$$\sum F_y = 0 : A_y + D_y - P = 0$$

(11.26)

Since the member BD is at a 1:1 slope, the vertical reaction (D_y) should be equal in magnitude to the horizontal reaction (D_x) at point D. Summing moments about point A confirms this assumption:

$$\sum M_A = 0 : PL - D_y \frac{L}{2} = 0 \rightarrow D_y = 2P$$

(11.27)

The positive result for D_y confirms the assumptions regarding direction in the free-body diagram.

Equilibrium conditions for member BD confirm the prior results as follows:

$$\sum F_x = 0 : B_x + D_x = 0 \rightarrow B_x = -D_x = -2P$$

$$\sum F_y = 0 : B_y - D_y = 0 \rightarrow B_y = D_y = 2P \qquad (11.28)$$

$$\sum M_B = 0 : D_y \frac{L}{2} - D_x \frac{L}{2} = 0 \rightarrow D_x = D_y$$

As shown, the directions of D_x and D_y are opposite for the member BD diagram as compared to the member AC diagram. This orientation follows from Newton's third law. The total magnitude of the tension load in member BD is as follows:

$$D_t = \sqrt{D_x^2 + D_y^2} = 2\sqrt{3}P \qquad (11.29)$$

The actual direction of the vertical reaction at point A (A_y) should be pointed downward, while the horizontal reaction at point B (B_x) should be pointed to the left. This frame can be modified to reflect different lengths and angles of the individual members. For example, if segment DC is doubled in length to L, the horizontal reactions at points A and B are increased to $3P$. For this case, the vertical reaction at point B would be increased to $3P$, while the downward-facing vertical reaction at point B would be $2P$. Additionally, the total tension load on member BD would be increased to $2\sqrt{2}P$.

Distributed Forces and Properties of Areas

From the study of structures and beams, it is apparent that many forces are actually distributed over the affected area or volume of the structure. As an example, consider the effect of gravity on a 2D or 3D structure shown in Figure 11.15.

We may consider each subelement of the structures as individual particles of infinitesimal size that are affected by the gravitational attraction to the center of the earth. However, since the distance from the subelements to each other is small compared to the distance to the earth's center, then we can locate the center of gravity of the structure and use it in our analysis. As an example, consider a beam structure that may represent a human bone such as a humerus as shown in Figure 11.16.

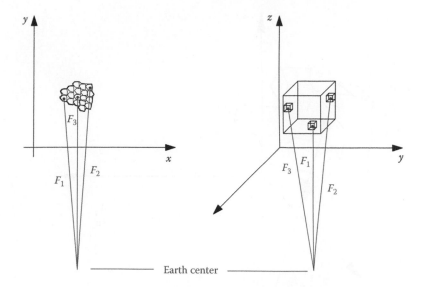

Figure 11.15 Two- and three-dimensional distributed forces.

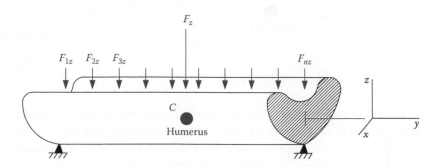

Figure 11.16 Symmetrical humerus beam representation.

Note that for simplicity, we have considered the distributed forces along the y-axis directed downward in the z-direction where they are given by

$$F_z = F_{1z} + F_{2z} + \cdots + F_{nz} = \sum_1^k F_n \tag{11.30}$$

$$\sum F_x = \sum F_y = 0 \tag{11.31}$$

For solid structures or structures that are geometrically contained, we may define the center of gravity, which is known as the first moment of inertia. Additionally, we may define the second moment of inertia of an area and the

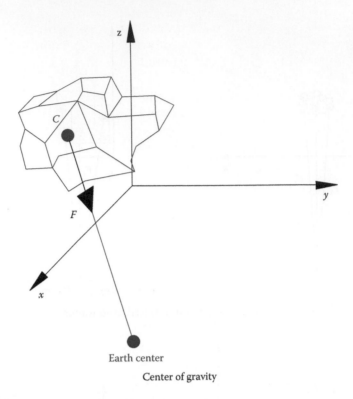

Figure 11.17 Centroid or center of mass.

mass moment of inertia. This type of analysis is required when investigating the properties of biological structures because we usually use the weight of the structure as a single force located at the centroid. For the 3D object shown in Figure 11.17, we define the centroid of the volume as

$$\bar{x}A_x = \int x\, dA_x$$

$$\bar{y}A_y = \int y\, dA_y \qquad (11.32)$$

$$\bar{z}A_z = \int z\, dA_z$$

This analysis lends itself to finding the location of the resultant weight of a load that may be distributed on a particular biological material. The preceding discussion may be used to analyze the distributed forces over the corresponding areas associated with them. The resultant of these forces can be obtained by summing the corresponding areas, and the moment of the resultant about any axis can be determined by computing the first moment of the area about the axis. It is also necessary for computational purposes to consider distributed forces whose magnitude depends on the element of the area and the distance to the axis.

The first moments are determined about a particular axis. Consequently, for 2D objects, we define the first moments as

$$Q_x = \int y \, dA \tag{11.33}$$

$$Q_y = \int x \, dA \tag{11.34}$$

Note that we have restricted these first moments to two dimensions in order to simplify the equations. For 3D objects, the equations are somewhat more complex and more difficult to visualize. Fortunately, for biomechanical calculations, most biological structures can be modeled in two dimensions so that the third dimension need not be considered by an appropriate selection of the axes. These axes are normally referred to as the principal axes. A generalized analysis of moments of inertia about arbitrary axes requires tensor analysis that includes nine components in order to solve the equations. Consequently, that type of analysis is beyond the scope of the book and simply not warranted in this context.

The second moments of inertia with respect to the particular axes are given as

$$I_x = \int y^2 \, dA \tag{11.35}$$

$$I_y = \int x^2 \, dA \tag{11.36}$$

and the polar moment of inertia is as follows:

$$J_0 = I_x + I_y = \int r^2 \, dA \tag{11.37}$$

The radius of gyration for these moments is defined as

$$k_x = \sqrt{\frac{I_x}{A}}; \quad k_y = \sqrt{\frac{I_y}{A}}; \quad k_0 = \sqrt{\frac{J_0}{A}} \tag{11.38}$$

Finally, for a 3D body, the moment of inertia can be obtained from the following equation:

$$I_v = \int r^2 dm = \rho \int r^2 dV \qquad (11.39)$$

where
    ρ is the density of the body
    m is the mass
    V is the volume
    r is the radial distance to the center of mass

Note the distinction of Equation 11.39 in that it represents a solid body rotating about an axis external to the body and not through one of the principal axes. We have tabulated the properties of common areas encountered in biomechanics, and these are shown in Figure 11.18.

Particle Kinematics

Mechanics when applied to biomechanical studies and calculations deals with the relationship between forces, bodies, and motion. The forces are produced by the combination of the relative motion of objects, the bodies, or of systems. The method used to describe motion from a mathematical standpoint is called kinematics or simply dynamics.

The motion of a particular body or body part is defined as its continuous change in position relative to an arbitrary coordinate system. For most applications in biomechanical calculations, we may use either a Cartesian, cylindrical, or spherical coordinate system. In any of these coordinate systems, the position of the body is specified by its projection into the three axes of a rectangular coordinate system. As the particular body moves along a path, the projections of the path to the respective axes move in straight lines. Thus, the motion of the body is reconstructed from the motions produced by the projections on the coordinate system. The relationship between these three systems is described by Figure 11.19.

The interrelationship between the systems is described for a point P in Cartesian coordinates as

$$P(x,y,z) \qquad (11.40)$$

where in cylindrical coordinates

$$x = \rho \cos \varphi \qquad (11.41)$$

Shape	Area	Area Moment
Circle a	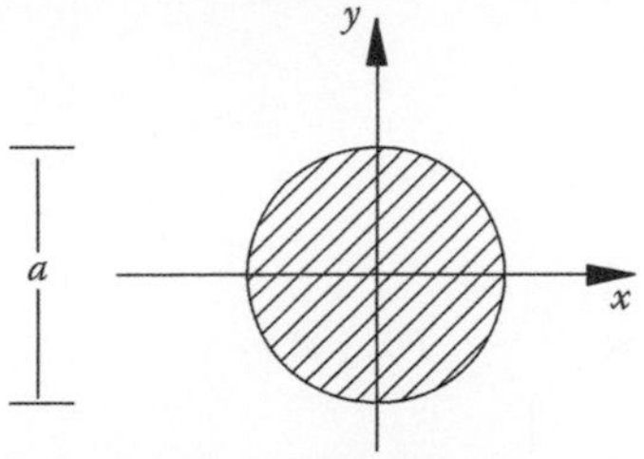 $A=\dfrac{\pi a^2}{4}$	$I=\dfrac{\pi a^4}{64}$
Ellipse a	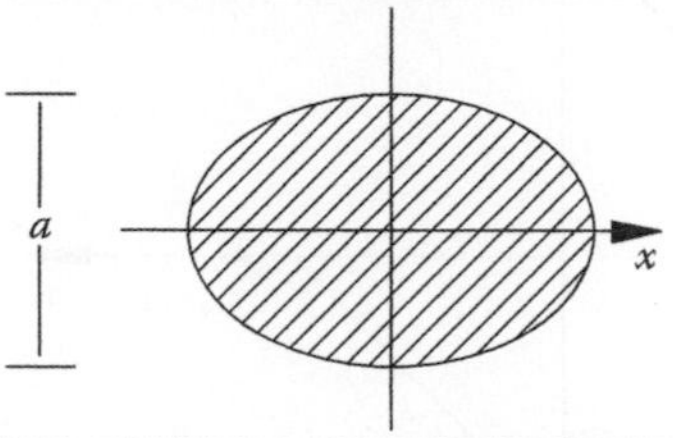 $A=\dfrac{\pi ac}{4}$	$I_x=\dfrac{\pi a^3 c}{64}$ $I_y=\dfrac{\pi ac^3}{64}$
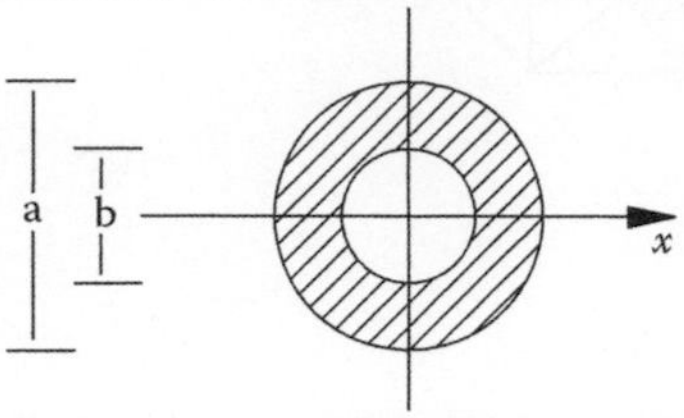 Hollow circle a b	$A=\dfrac{\pi(a^2-b^2)}{4}$	$I=\dfrac{\pi(a^4-b^4)}{64}$
Hollow ellipse a b	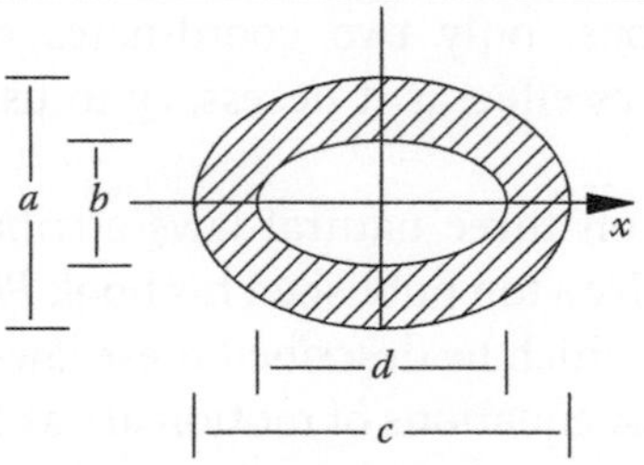 $A=\dfrac{\pi(ac-bd)}{4}$	$I_x=\dfrac{\pi(a^3c-b^3d)}{64}$ $I_y=\dfrac{\pi(ac^3-bd^3)}{64}$

Figure 11.18 Properties of areas.

$$y = \rho\sin\varphi \tag{11.42}$$

$$z = z \tag{11.43}$$

and in spherical coordinates

$$x = r\sin\theta\cos\varphi \tag{11.44}$$

$$y = r\sin\theta\sin\varphi \tag{11.45}$$

$$z = r\cos\theta \tag{11.46}$$

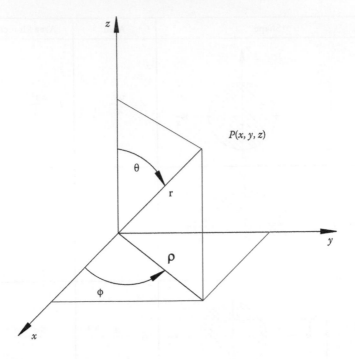

Figure 11.19 Coordinate systems.

For most biomechanical applications, only two coordinates need to be considered. Only when rotation takes effect is it necessary to use the third coordinates.

Classical mechanics are based on three natural laws attributed to Sir Isaac Newton (1643–1727). In 1686, Newton published his book *Philosophiae Naturalis Principia Mathematica* in which he described these laws. Based on an inertial reference system, the basic equations of motion are as follows:

$$v = \frac{dx}{dt}, \quad a = \frac{dv}{dt} \tag{11.47}$$

where
  v is the velocity
  a is the acceleration
  x is the displacement
  t is the time

For uniformly accelerated linear motion, it can be shown that

$$x = x_0 + v_0 t + \frac{1}{2} a t^2 \tag{11.48}$$

$$v = v_0 + at \qquad (11.49)$$

For uniformly accelerated curvilinear motion, the angular velocity ω, the angular acceleration $\forall$, and the angular displacement θ are governed by

$$\omega = d\theta/dt; \quad \alpha = d\omega/dt \qquad (11.50)$$

and

$$\omega = \omega_0 + \alpha t \qquad (11.51)$$

$$\theta = \theta_0 + \omega_0 t + \frac{1}{2}\alpha t^2 \qquad (11.52)$$

In some biomechanical calculations, it may be necessary to relate angular to linear velocity and acceleration. Figure 11.20 allows for the determination from the basic equations:

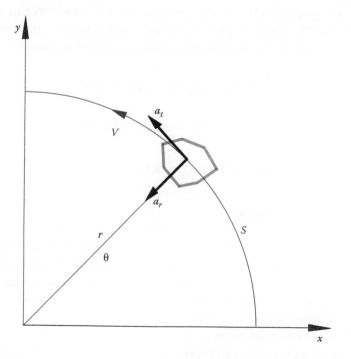

Figure 11.20 Angular and linear movement.

The arc produced by a body is given by

$$s = r\theta \tag{11.53}$$

where
    s is the arc length
    θ is the angular displacement
    r is the radial distance from the reference origin

The velocity is

$$v = r\omega \tag{11.54}$$

and the acceleration has two components, one in the tangential direction and the other in the radial direction,

$$a = r\alpha a_t + \frac{v^2}{r} a_r \tag{11.55}$$

where a_t and a_r are unit vectors in the tangential and radial directions.

At the beginning of this chapter, we discussed Newton's laws in an introductory form. The first law is self-explanatory and generally not expressed in the equation form. The second and third laws are expressed in equation form as

$$F = ma \quad \text{Newton's Second Law} \tag{11.56}$$

$$F = -F' \quad \text{Newton's Third Law} \tag{11.57}$$

Bolded symbols represent vector quantities.

Conservation of Mass

One of the most fundamental principles of physics is that of the conservation of mass, which is consistent with the principles of conservation of energy and conservation of momentum. This principle simply states that within a closed system, mass is neither created nor destroyed. Mass conservation applies to a variety of areas. Mass is conserved in chemical processes, in mechanics, and in fluid dynamics. In biomechanics, mass is conserved unless the system becomes open such as when an individual bleeds or is shot and mass is added in the form of a bullet fragment.

In the context of this book, we can rightly assume that the mass of a body or system can be determined by simply multiplying the volume, V, of the object by its density, ρ. Thus, for mass conservation, the product of the volume and the density must remain constant:

$$m = \frac{w}{g} = \rho V \tag{11.58}$$

Also recall that mass is the ratio of weight, w, divided by gravity, g. In a more general sense, the conservation laws that are applied in physics include mass, energy, electric charge, heat, and momentum. Conservation laws are described by continuity equations for the particular phenomenon of the transport of the quantity that is conserved. For biomechanical analysis, we are more concerned with the conservation of energy and conservation of momentum whose sections follow.

Conservation of Momentum

In order to consider the concept of momentum we refer back to Newton. Newton's laws of motion give rise to the concepts of work and energy, which lead to the concepts of impulse and momentum. We begin by first considering impulse and momentum. In the next section, we deal with work and energy. Both concepts are similarly derived from Newton's second law.

$$F = ma = m\frac{dv}{dt} \tag{11.59}$$

Rearranging Equation 11.59 and providing limits of integration, we obtain

$$\int_{t_1}^{t_2} F\, dt = \int_{v_1}^{v_2} m\, dv \tag{11.60}$$

The quantity on the left side of Equation 11.60 is called the impulse of the force F over the time interval $t_2 - t_1$. Thus, by definition

$$\text{IMPULSE} = \int_{t_1}^{t_2} F\, dt \tag{11.61}$$

The impulse is a vector quantity that can only be evaluated when the force F is known as a function of time. In biomechanical analysis of vehicular impacts, the time element is not generally known but can be approximated, which is especially true in some instances where test measurements are performed with accelerometers. As a rough estimation, the impact pulses are approximately 100 ms but may vary between 50 and 150 ms. In many instances, the force may be calculated based on the results of the right side of Equation 11.60:

$$\int_{t_1}^{t_2} F\,dt = mv_2 - mv_1 \tag{11.62}$$

Equation 11.62 expresses a subtle yet very important concept that applies to objects that strike each other, such as when bodies fall or are struck by other objects. The significance of this equation states that the "vector impulse" of the resultant force on a particle or object, over the time interval, is equal in magnitude and direction to the "vector change" in momentum of the particle or object. The impulse momentum principle is chiefly applied to short duration forces arising from explosive events or impacts. These types of forces are often referred to as impulsive forces. Such forces are imparted when human body parts are struck by objects such as bats, clubs, or bullets.

In order to apply conservation of momentum to a particular problem, we note that momentum is always conserved. That is, the momentum prior to impact must be equal to the momentum after the impact. In equation form, the initial and final momentums are expressed as

$$M_i = M_f \quad \text{or} \quad (mV)_i = (mV)_f \tag{11.63}$$

where
M_i is the initial momentum
M_f is the final momentum

For two bodies or objects Equation 11.63 becomes

$$m_1 V_{1i} + m_2 V_{2i} = m_1 V_{1f} + m_2 V_{2f} \tag{11.64}$$

where m and V are the respective masses and velocities.

Impacts between two bodies may be classified as elastic, inelastic, or somewhat between the two. A completely elastic impact is one in which the two bodies come together and then separate. In an elastic impact, the forces of interaction between the two bodies are conserved, and the total kinetic energy is the same before and after impact. Completely elastic impacts only occur when two very solid objects impact one another as between two billiard balls. When a bat or club strikes a human body part, the impact is not completely elastic because of the cushioning effect of the skin, muscle, and fat surrounding the struck area. However, a reasonable approximation of an elastic impact can often be made in such events, especially if the head or a bony portion of the anatomy is struck.

The opposite of an elastic impact is when an impact occurs between two objects and then the two objects move in unison. This impact is classified as inelastic. Again, it should be emphasized that no impacts are completely inelastic because relative movement between the two objects always occurs. The closest approximation to an inelastic impact may be a gunshot when the bullet is lodged in the body. If the bullet passes through the body, then the impact would not be completely inelastic. In other words, some restitution of momentum would occur. Restitution in human injuries is difficult to quantify so that generally we either assume, as an approximation, that the impact was either elastic or inelastic, depending on the specifics of the impact. If one of the bodies, say a person, is at rest initially and is struck with a bat, $V_{2i} = 0$, Equation 11.64 reduces to

$$V_{1i} = V_{1f} + \frac{w_2}{w_1} V_{2f} \tag{11.65}$$

If we further simplify by noting that a striking bat or bullet would have lost all its speed upon impact and that the mass of the individual is much greater than the mass of the bat or bullet, then the speed of the striking object would be as follows:

$$V_{1i} \cong \frac{w_2}{w_1} V_{2f} \tag{11.66}$$

Once the speed is known, the acceleration can be calculated, and the force can be computed, which allows for a determination if the biomechanical injury is possible. For a gunshot, the initial speed of the bullet can simply be determined from the muzzle velocity of the charge in the shell. This data is readily available. If the bullet passes through the body and exits, the use of Equation 11.66 does not give accurate results for V_{2f}.

Conservation of momentum is extremely fundamental and powerful. It is more fundamental and general than the principle of conservation of mechanical energy. For most applications in mechanics, momentum techniques have a wider application than energy methods. However, for biomechanical calculations, this statement is not necessarily true because of the fact that restitution is very difficult to estimate with a good degree of confidence. Momentum holds true no matter what the nature of the internal forces acting on the bodies or particles may be. In contrast, mechanical energy is conserved only when the internal forces are themselves conservative. We now turn our attention to energy and work and discuss their limitations in biomechanics.

Conservation of Energy

In order to consider how energy and work affect bodies or particles in motion, we consider the diagram shown in Figure 11.21 representing a hammer blow. Note that both translation and rotation occur in the motion of the hammer.

The motion of the hammer head is a curved path representing the trajectory of the hammer head of mass m moving in the x–y plane acted upon by a force F. The force is produced by an assailant as he swings the hammer to produce injury. We choose to resolve the force onto its components F_s and F_n along the path. The normal component of the force F_n is the centripetal force and its effect is to change the direction of the velocity v. The effect of the component F_s is to change the magnitude of the velocity. Obviously, as the hammer arcs down, it will not only pick up energy due to its motion but also by its relative position due to gravity. Thus, we can conclude that two types of energy are produced when we

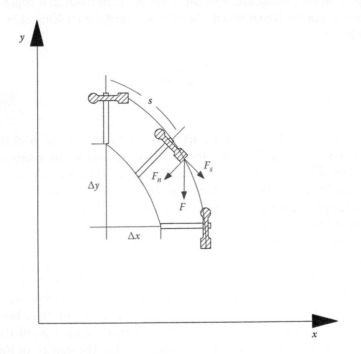

Figure 11.21 Bodies in motion.

discuss conservation of energy. These are kinetic and potential energy. We will look at each separately.

Kinetic Energy

In general terms, the magnitude of F_s will be a function of the path s, where s is the distance that the hammer head has traveled from a fixed reference. Applying Newton's second law

$$F_s = ma = m\frac{dv}{dt} \tag{11.67}$$

Since F_s is a function of the path s, we may apply the chain rule

$$\frac{dv}{dt} = \frac{dv}{ds}\frac{ds}{dt} = v\frac{dv}{ds} \tag{11.68}$$

$$F_s = mv\frac{dv}{ds} \tag{11.69}$$

If v_1 is the velocity when $s = s_1$ and v_2 is the velocity when $s = s_2$, we may integrate Equation 11.69:

$$\int_{s_1}^{s_2} F_s\,ds = \int_{v_1}^{v_2} mv\,dv \tag{11.70}$$

The integral on the left side of Equation 11.70 is the work W of the force F_s and is defined as

$$W = \int_{s_1}^{s_2} F_s\,ds \tag{11.71}$$

The integral given by Equation 11.71 can only be integrated when F_s is known as a function of s or when F_s and s are related by another variable. The integral on the right side can always be evaluated:

$$\int_{v_1}^{v_2} mvd = \frac{1}{2}mv_2^2 - \frac{1}{2}mv_1^2 \tag{11.72}$$

Equation 11.72 represents the kinetic energy of the body or particle and is represented as

$$E_k = \frac{1}{2}mv^2 \qquad (11.73)$$

Equation 11.73 is generally written in terms of the work W as

$$W = E_{k2} - E_{k1} \qquad (11.74)$$

stating that the work of the resultant force exerted on the body equals the change in the kinetic energy and is known as the work–energy principle.

In order to properly apply the work and the energy in a biomechanical computation, some basic explanations are necessary. Work on an object or particle is only done when the forces are exerted while the body moves along the line of motion. If the component of the force is in the same direction as the displacement, the work is positive. If the force is opposite to the displacement, the work is negative. If the force is perpendicular to the displacement, the work is zero.

When we compute the work of a force, we multiply the magnitude of the vector F in the direction of the vector ds. This is the scalar product of the vectors so that

$$W = \int_{s_1}^{s_2} F \cdot ds = \int_{s_1}^{s_2} F \cos\theta \, ds \qquad (11.75)$$

In the special case when $\cos\theta = \pm 1$,

$$W = \pm Fs \qquad (11.76)$$

Only in this special case is the work equal to the force times the distance and that only the total work is equal to the change in the kinetic energy of the body or particle. Kinetic energy and work are scalar quantities. The magnitude of the velocity of a moving object is the only parameter that establishes the amount of kinetic energy. Neither the process that produced the motion nor the direction of motion establishes the value of the kinetic energy. The work–energy principle states that the change in the kinetic energy does not depend on the individual values of the force F and the path s. Kinetic energy increases if the work is positive and decreases if the work is negative. A particle or body traveling at a constant velocity does no work and the change in the kinetic energy is zero.

Potential Energy

Bodies or particles in motion, whether they are humans, tools, weapons, bullets, or equipment, are near the surface of the earth so that variations in the gravitational force are neglected. Figure 11.22 represents a body moving vertically along an arbitrary path.

The downward force on the body is produced by its weight w, and $\boldsymbol{P}$ is the resultant of all the other forces acting on the body. The work of the gravitational force is given as

$$W_g = \int_{s_1}^{s_2} w \cos\theta\, ds \tag{11.77}$$

Since N is the angle between ds and the vertical component dy, then $dy = ds \cos N$, and $dy = -\cos\theta\, ds$, thus

$$W_g = -\int_{y_1}^{y_2} w\, dy = w(y_2 - y_1) = (mgy_2 - mgy_1) \tag{11.78}$$

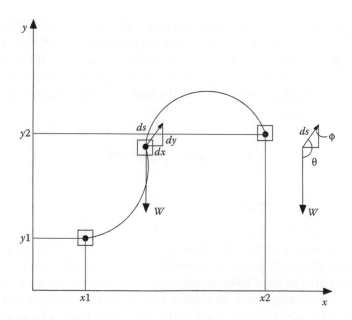

Figure 11.22 Gravitational potential energy.

Therefore, the work of the gravitational force depends only on the initial and final elevations and not the path. Since the total work is equal to the total change in the kinetic energy, we have

$$W_t = W_p + W_g = E_{k2} - E_{k1} \tag{11.79}$$

where
   W_t is the total work
   W_p is the work of the force $\mathbf{P}$
   W_g is the work due to gravitational effects
   E_{k2} is the final kinetic energy
   E_{k1} is the initial kinetic energy

Generally, it is convenient to express Equation 11.79 in terms of the work of the force $\mathbf{P}$ so that

$$W_p = \left(\frac{1}{2}mv_2^2 + mgy_2\right) + \left(\frac{1}{2}mv_1^2 + mgy_1\right) \tag{11.80}$$

where the gravitational potential energy is given as

$$E_p = mgy \tag{11.81}$$

The sum of the kinetic and potential energy is called the total mechanical energy. The work of all the forces acting on the body, except for the gravitational force, equals the change in the mechanical energy. If the work W_p is positive, the mechanical energy increases; if negative, it decreases. In the special case in which the only force on the body is the gravitational force, the work W_p is zero, then

$$\frac{1}{2}mv_2^2 + mgy_2 = \frac{1}{2}mv_1^2 + mgy_1 \tag{11.82}$$

Elastic Potential Energy

Another concept that may be used in biomechanical calculations is that of elastic potential energy. This principle is introduced in the modeling of compression of a body part and rebound produced by the biological structures. It may also be used when modeling elastic tissues such as tendons and ligaments. Essentially, these structures may behave as springs and behave according to Hooke's law, which states that the elastic force is proportional to the displacement of the spring as displayed in Figure 11.23.

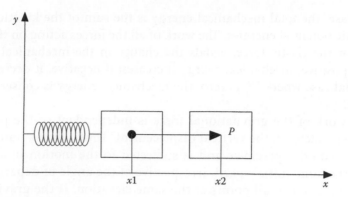

Figure 11.23 Hooke's law.

When an external force P stretches or compresses the spring so as not to deform it, the elastic force is given by

$$F = kx \tag{11.83}$$

where
   k is the stiffness coefficient of the spring structure
   x is the displacement

The work of the elastic force is given by

$$W_e = \int F \cdot ds = \int_{x_1}^{x_2} F \cos\theta \, dx \tag{11.84}$$

Since the force F is opposite, the direction of dx, $\cos\theta = -1$ and

$$W_e = -\int_{x_1}^{x_2} kx \, dx = \left(\frac{1}{2}kx_1^2 - \frac{1}{2}kx_2^2 \right) \tag{11.85}$$

As in the previous section, we let W_p be the work of the applied force P so that the total work is equal to the change in the kinetic energy and

$$W_t = W_p + W_e = E_{k2} - E_{k1} \tag{11.86}$$

The term $(1/2)kx^2$ represents the elastic potential energy. We may express Equation 11.86 in a similar fashion as Equation 11.80:

$$W_p = \left(\frac{1}{2}mv_2^2 + \frac{1}{2}kx_2^2 \right) - \left(\frac{1}{2}mv_1^2 + \frac{1}{2}kx_1^2 \right) \tag{11.87}$$

n this case, the total mechanical energy is the sum of the kinetic energy
nd elastic potential energies. The work of all the forces acting on the body,
xcept for the elastic force, equals the change in the mechanical energy.
W_p is positive, mechanical energy increases; if negative, it decreases. In
special case where W_p is zero, the mechanical energy is conserved and
constant.

The work of the gravitational force is independent of the path and
purely dependent on the vertical displacement. Thus, when a human falls
or is pushed off a precipice and is subjected to the motion of a projec-
tile, for the same initial speed, irrespective of the angle of departure, the
speed is the same at all points at the same elevation. If the gravitational
force acts alone on the body, the total mechanical energy is conserved. If
the body first rises and then descends to its original position, the work
is completely recovered. Similarly, in the extension and contraction of a
spring or tendon to the original position, the elastic potential energies are
conserved, and the work is recovered. That is not the case if the tendon
extension exceeds the elastic limit, and the tendon is overstretched or ruptured. Thus, conservative forces are characterized by the independence of
path, equality in the difference between initial and final energy functions,
and are completely recoverable. That is not the case when the forces are
dissipative as when friction occurs in that the path plays a part in the dis-
sipation of the forces.

In a more general sense, both internal and external forces produce a
change of the kinetic energy of a body or system, which is represented by the
following equation:

$$W_t = W_o + W_i = \Delta E_k \tag{11.88}$$

A more general representation for the work produced by internal and external forces is then

$$W_p = W_t + \Delta E_k = W_t + \Delta E_{ok} + \Delta E_{ik} \tag{11.89}$$

where
   W_e is the work of the elastic force
   W_p is the work of the force P
   W_t is the total work = $W_p + W_e = \Delta E_k$
   W_o is the external work
   W_i is the internal work
   ΔE_k is the kinetic energy change
   ΔE_{ok} is the external potential energy change
   ΔE_{ik} is the internal potential energy change

Vibration: Whiplash Models

Two of the most serious biomechanical calculations that are performed involve shaken baby syndrome and whiplash. The violent nature of shaken baby syndrome yields a method of calculation to determine the injury effect of this serious crime. At the other end of the spectrum, whiplash is probably the most abused form of injury claim and results in significant monetary consequences for all of us resulting from increased insurance premiums. Both of these injury mechanisms may be studied from a vibration perspective. In both of these effects, the head moves back and forth simulating a vibrating string motion. For the shaken baby syndrome, the repetitive cyclic action may last for a few seconds as the baby's body moves back and forth while the head attempts to follow creating significant strain on the neck and the brain. The motion of the brain leads to closed head injury and may result in death. For the whiplash effect, the motion is that of a damped sinusoid with a peak value as the head is thrust rearward and then bounces forward. The duration of this pulse is in the neighborhood of 150 ms. The model is represented by Figure 11.24.

For this model, the circle represents the head, and the neck is represented by a cantilever beam similar to a diving board. The force P is related to the displacement x and the spring constant k by

$$P = kx \qquad (11.90)$$

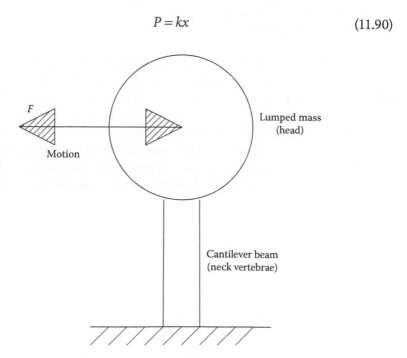

Figure 11.24 Lumped mass cantilever model.

Summing forces, it can be shown that

$$m\frac{d^2x}{dt^2} + kx = F\sin\theta t = P \tag{11.91}$$

where θ is the angular velocity of the forcing function. The displacement is given by

$$x = \frac{F\sin\theta t}{m(w^2 - \theta^2)}; \quad w^2 = \frac{k}{m} \tag{11.92}$$

The maximum value of the displacement is given as

$$x_{max} = \frac{P}{(k - m\theta^2)} = \frac{P}{((3EI/l^3) - m\theta^2)} \tag{11.93}$$

where
  E is Young's modulus for the material
  $I = (\pi d^4/64)$ represents the moment of inertia of the neck beam with diameter d
  m is the mass of the head
  l is the vertebral length

For computations involving whiplash, the ultimate strength of disc cartilage may vary from approximately 0.2 to 0.32 kg/mm². For shaken baby syndrome, the shearing properties of cerebral dura mater are known for adults and children beyond the age of 10. For 10–19 year olds, the shearing breaking load is approximately 1.1 kg/mm, and the ultimate shearing strength is approximately 2.32 kg/mm². Shaken baby syndrome occurs generally for children who are less than 1 year old. Consequently, the values for the shearing breaking load and ultimate shearing strength should be considered as absolute maximums.

Errors, Sensitivity, Uncertainty, and Probability

12

Misconceptions

Over the past few years, a great emphasis has been placed on errors in the forensic sciences. This emphasis has come about as a result of the 1993 Supreme Court Dauber decision, the Kumho Tire case, and the National Academy of Sciences report in 2010 "Strengthening Forensic Science in the United States: A Path Forward." In this report, among other findings two stand out with respect to the title of the chapter and these are as follows:

1. *The small quantity or number of scientific research used to confirm the validity and reliability of forensic disciplines and established quantifiable measures of uncertainty in the conclusions of forensic analysis.*
2. *The small number of research programs on human observed bias and sources of human error in forensic examinations.*

This paucity, or incompleteness of data, or error, is often used to attack an expert in a manner that implies that data has either been skewed, disregarded, falsified, or that the computations are outside of standard scientific methodology and are therefore erroneous. Although many of the criticisms are warranted, opposing attorneys have mischaracterized the rulings and the cases in order to attack the opposing expert. Many jurisdictions have now supplanted the federal rules of evidence with Daubert style hearings on the admissibility of the evidence provided by experts. One phalanx of attack is to claim that the computations are full of errors so that they fall outside of the recognized values or procedures. Another method of attack is to state that all possible variations and conditions have not been properly investigated. Recall the trillions of permutations alluded to in the discussion of Chapter 8. We see that type of attack on experts frequently in criminal as well as civil cases. When a suspect is apprehended and the evidence is very pointed to that individual, the suspect may be arraigned and charged and brought to trial. A typical defense includes allegations that all suspects were not excluded so that the suspect is wrongly accused. In other words, the investigating officers did not question all possible suspects whether they existed or not.

Consider the following example from a real case in a civil trial. A lady in a sedan claims that her vehicle was rear-ended by an SUV that then fled the scene. Inspection of the damage to her vehicle is not consistent with a rear-end collision. In fact, the damage is consistent with a sharp horizontal object being dropped on the trunk and rear bumper of the sedan. As a result of the alleged collision, the lady is claiming a multitude of ailments, including diseased teeth requiring extraction and dental implants. The insurance company denies the claim because of the obvious discrepancies in the damage and the claimed injuries. Before trial, the attorney for the plaintiff makes a motion for a Daubert hearing to disqualify the expert because the expert did not include all vehicles ever made since the inception of the automobile. The expert did consider all SUVs but not, for example, a 1922 Cadillac. Notwithstanding the fact that SUVs have only been manufactured in the last couple of decades, and are outside of the boundaries being claimed, the expert is not bound to consider data not relevant such as a Martian Space Ship having caused the damage. Needless to say, the Daubert hearing did not disqualify the expert. Keep this in mind as a later example will further expound upon the arguments made.

Error

Before we begin the mathematical treatment of this chapter, we need to expand on the terms used to define the uncertainty in forensic calculations. We begin with error. Error is defined in the dictionary and in people's minds as a mistake, inaccuracy, miscalculation, blunder, or oversight. It is the state or condition of being wrong in conduct or judgment. An argument is generally made that the techniques, equations, data, or some other form of analysis is incorrect or outside of the realm of accepted practice. When error or error rate is calculated or implied in the sciences, it has a totally different meaning. Consider the following:

> *Statistical error is not a mistake but rather the difference between the computed, measured, or estimated value and the actual value.*
> *In mechanics, when a system of the form $Ax = b$ is solved, there is a difference in the inaccuracy in x and the residual, the inaccuracy in Ax.*
> *In mathematics, error is the difference between an actual value and the estimate or computed value. Error also means the difference between the actual value and the truncated value where only the first few terms of an infinite series are used.*

A better description of error is then the uncertainty in the computation or assessment of the values used in the computation. In nature, many parameters vary within some specified range. As an example, the strength of bone is known within a range of values, which does not mean that when performing a calculation on a bony structure, we may substitute any value for its strength or utilize a value outside of the know range of the parameter. Therefore, it is better to state the variability of the parameter within the known uncertainty and perform the calculations accordingly. Simply put, we should give the range of values that include the uncertainty and the variability.

The arguments posed earlier lead to another term called sensitivity. In our mathematical formulation of the models that we use to describe an effect (the equations we use to make the computations), some parameters that vary will have a great effect on the outcome. Others may not affect the computations as much or maybe not at all. This analysis or measure is what we call sensitivity.

When being cross-examined, it behooves the forensic practitioner to clearly and succinctly explain what is meant by error in a scientific or mathematical context. It is best to answer in terms of sensitivity, uncertainty, and probability instead of error so as to clarify the concepts for the judge and the jury. Assuming that the computations have been carried out correctly, when asked to state the error, it is much more descriptive to say that there is no error, but there is variability and uncertainty within certain bounds dictated by the known scientific literature and the scientific method. By performing a variety of computations for all the variables that are sensitive in the equations, and those that are not, the essential constituents are analyzed. It is very helpful to graph or plot the results as these make strong demonstrative evidence in the courtroom or in a deposition. Many of the terms in an equation may not be subject to change as they could be constants such as the acceleration of gravity. In general, most equations in science may be functions of several variables.

Sensitivity

In general, an equation dependent on n variables can be expressed as

$$P = f(x_1, x_2, \ldots, x_n).$$ (12.1)

First, let us consider the case where an equation is a function of only one variable $P = f(x)$.

The sensitivity of P to the variable x is defined as

$$S_x^P = \lim_{\Delta x \to 0} \frac{\Delta P/P}{\Delta x/x} = \frac{x}{P} \lim_{\Delta x \to 0} \frac{\Delta P}{\Delta x} = \frac{x}{P} \frac{\partial P}{\partial x} = \frac{\partial(\ln P)}{\partial(\ln x)}.$$ (12.2)

Applying the basic definition of sensitivity, the following relationships may be developed:

$$S_x^{cx} = 1; \quad c = \text{constant} \tag{12.3}$$

$$S_x^P = -S_x^{1/P} \tag{12.4}$$

$$S_x^P = -S_{1/x}^P \tag{12.5}$$

$$S_x^{P1P2} = S_x^{P1} + S_x^{P2} \tag{12.6}$$

$$S_x^{P1/P2} = S_x^{P1} - S_x^{P2} \tag{12.7}$$

$$S_{x^n}^P = \frac{1}{n} S_x^P \tag{12.8}$$

$$S_x^{P1+P2} = \frac{P1 S_x^{P1}}{P1 + P2} + \frac{P2 S_x^{P2}}{P1 + P2} \tag{12.9}$$

Since most functions are in terms of various parameters, it is important to determine the parameter deviations. To do so, we begin with the basic definition as expressed in Equation 12.2. For small deviations in x, we can determine the change in the function due to one element as

$$\nabla P \cong S_x^P \frac{\Delta x}{x} P. \tag{12.10}$$

For multiple elements, we may expand Equation 12.9 in a Taylor series to obtain

$$\Delta P \cong \frac{\partial P}{\partial x_1} \Delta x_1 + \frac{\partial P}{\partial x_2} \Delta x_2 + \cdots + \frac{\partial P}{\partial x_m} \Delta x_m + \text{higher order terms.} \tag{12.11}$$

For small variations in Δx_j, we neglect the higher order terms so that

$$\Delta P \cong \sum_{x_j}^{m} \frac{\partial P}{\partial x_j} \Delta x_j = \sum_{x_j}^{m} \left[\frac{\partial P}{\partial x_j} \right] \left[\frac{x_j}{P} \right] \left[\frac{\Delta x_j}{x_j} \right] P = \sum_{x_j}^{m} S_{x_j}^P V_{x_j} \tag{12.12}$$

where

$$V_{x_j} = \frac{\Delta x_j}{x_j} = \text{per unit change in the parameter } x \text{ and is known as}$$

the variability of x.

For example, a simple but quite accurate approximation for the tensile stress on a human vertebra is given by

$$\sigma_s \cong \frac{0.003 w_T}{ndA} v^2 \qquad (12.13)$$

where
  n is the number of vertebrae
  d is the distance from the lower spine to the rest of the spine (approximately 0.2 ft)
  A is the cross-sectional area on the spine (approximately 660 mm²)
  w_T is the weight of the individual in pounds
  v is the speed of impact (in ft/s)

Applying Equation 12.7 to Equation 12.12, we find the following sensitivities:

$$S_{w_T}^{\sigma_s} = 1; \quad S_v^{\sigma_s} = 2; \quad S_n^{\sigma_s} = S_d^{\sigma_s} = S_A^{\sigma_s} = -1 \qquad (12.14)$$

$$\frac{\Delta \sigma_s}{\sigma_s} = S_{w_T}^{\sigma_s}\left[\frac{\Delta w_T}{w_T}\right] + S_v^{\sigma_s}\left[\frac{\Delta v}{v}\right] + S_n^{\sigma_s}\left[\frac{\Delta n}{n}\right] + S_d^{\sigma_s}\left[\frac{\Delta d}{d}\right] + S_A^{\sigma_s}\left[\frac{\Delta A}{A}\right] \qquad (12.15)$$

$$\frac{\Delta \sigma_s}{\sigma_s} = \frac{\Delta w_T}{w_T} + \frac{2\Delta v}{v} - \frac{\Delta n}{n} - \frac{\Delta d}{d} - \frac{\Delta A}{A} \qquad (12.16)$$

It is interesting to note that if all the parameters above vary by 10%, the tensile stress on the vertebrae does not change at all. This example shows how variability in one parameter may counteract the variability in another parameter. This example also shows how we may approach the variation in a calculation to show how a particular parameter affects the outcome of the calculation. For example, if we just consider the variation in the velocity of Equation 12.13 and let the velocity change from say 5 to 6 mph, a change of 20%, the change in the stress varies by an increasing factor of two. The stress equation is very sensitive to speed but not as sensitive to the other parameters.

Probability

As in all of science, when performing biomechanical calculations, it becomes necessary to perform some statistical computations. The computations are always based on a central measure of the parameter in question and are defined in this section. If we assume that the range of events varies from 1 to n so that for every value of x, there is a descriptive term $f_i \geq 0$. This descriptive term may be the frequency, the mass, the stress, the probability, or even the reliability of the occurrence of the event. The total weight is the sum of all the possibilities, or

$$N = \sum_i^m f_i. \tag{12.17}$$

The arithmetic mean is

$$x_{am} = \sum_i^m \frac{f_i x_i}{N}. \tag{12.18}$$

The geometric mean is

$$x_{gm} = \left[\prod_i^m x_i^{f_i} \right]^{1/N}. \tag{12.19}$$

The mode M_o for unweighted terms $(x_1, \ldots, x_N)$ is the value about which x_i most densely clusters. The median M_e for unweighted terms is the value equal or exceeded by exactly half of the values x_i. The root mean square is given by

$$RMS = \sqrt{\frac{\sum_i^N \left[f_i x_i^2 \right]}{N}}. \tag{12.20}$$

The standard deviation is

$$sd = \sqrt{\frac{\sum_i^N [x_i - x_m]^2}{N}}. \tag{12.21}$$

The variance is

$$V = [sd]^2.$$

(12.22)

The probability that an event E will occur as given by

$$P(E) = \frac{m}{n}$$

(12.23)

where
   E is the event
   P(E) is the probability that the event will occur
   m is the corresponding event
   n is the number of ways possible

If we consider two events A and B, the following relations hold true:

   P(A) is the probability of event A.
   P(B) is the probability of event B.
   P(A′) is the probability that event A does not occur.
   P(A/B) is the conditional probability of event B, given event A.
   P(A ∪ B) is the probability that event A and or event B occurs. Union.
   P(A ∩ B) is the probability that event A and event B both occur. Intersection.

$$P(A \cap B) = P(A)P(B)$$

(12.24)

Protective Structures and Their Effect

13

Panniculus adiposus is the medical term for adipose tissue, commonly referred to as fat. This fatty layer is superficial to the deeper layer of muscles of the human body. The muscular layers are known as the panniculus carnosus. The muscle layers include fascia by some references but not by others. Fascia is the connective tissue of fibers that attach and stabilize as well as separate and enclose muscles and organs. These flexible structures resist directional tension and provide a certain amount of protection to the human body. Consequently, the combination of adipose tissue, fascia, and muscles provides a level of protection to injury.

Chapter 8 outlined the biomechanical properties of muscles, bones, tendons, ligaments, and other structures that provide stability and, of course, a great level of protection to humans from injury. Recall the role and properties of intervertebral discs for example. In this chapter, we wish to identify two types of protective structures: those associated with the body, in particular fascia and adiposity, and other structures that are essentially man made and external to the body such as helmets, shoes, and clothing. We will not discuss the protective structures in automobiles such as seat belts, child restraint, seats, and airbag systems. These protective structures have been extensively analyzed and are specific to a device, the automobile, rather than more general. Our emphasis is on biological protection and simple man-made protective structures.

Fascia

Katake (1961) determined the tensile and expansive properties of human fascia from the thigh and leg of persons. These individuals were in the age group between 30 and 39 years old. Figure 13.1 shows the stress–strain curve in tension in the parallel direction to the fibers of fascia.

The tensile breaking load per unit width for fascia of the thigh was found to be 1.14 ± 0.17 and 1.04 ± 0.14 kg/mm for the leg fascia. The ultimate tensile strength in the parallel direction was found to be 1.47 ± 0.21 kg/mm^2 for

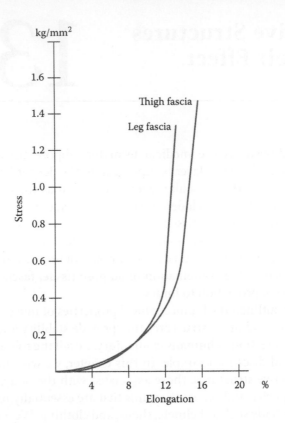

Figure 13.1 Stress–strain in tension of fascia.

the thigh and 1.31 ± 0.18 kg/mm² for the leg. The ultimate percent elonga-tion in a parallel direction was found to be 16.7% ± 0.52% for the thigh and 14.4% ± 0.51% for the leg.

In an oblique direction relative to the alignment of the fibers, these values were somewhat different. The tensile breaking load was about 33% of those in the parallel direction. The ultimate elongation in the oblique direction was approximately 1.2 times the elongation in the parallel direction. No sexual difference was found in these properties.

For the expansive properties, the following data was revealed by the stud-ies of Katake. The ultimate expansive strength of fascia was 53 ± 2.2 kg/cm² for the thigh and 52 ± 2.6 kg/cm² for the leg. The ultimate expansive strength per unit thickness was found to be 67.7 ± 1.2 kg/cm²/mm for the thigh and 67.6 ± 1.8 kg/cm²/mm for the leg. The ultimate expansion for a 7 mm area in diameter was found to be 0.08 mL for the thigh and 0.06 ± 0.001 mL for the leg. Again, no sexual differences were found for these properties. Figure 13.2 shows these properties.

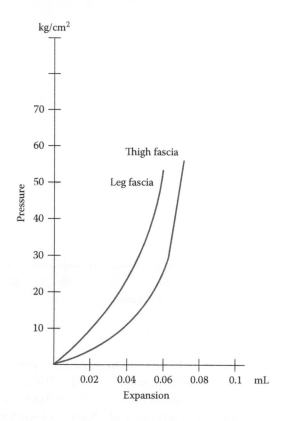

Figure 13.2 Pressure–expansion curves of fascia.

Panniculus Adiposus

No data is available for the properties on human adiposity. Consequently, we must rely on data from some limited animal studies. The density of fat is approximately 0.9 g/mL as compared to the density of muscle, which is about 1.06 g/mL. This simple fact explains why obese people will float better than muscular people. The reader may have noticed that swimmers are generally much leaner with respect to musculature than sprinters and weight-lifters. Simply, the more muscular the individual, the more effort is required to stay afloat. Consequently, more energy is expended in a vertical direction by the muscular swimmer rather than horizontal in propulsion through the water. This small difference in buoyancy characteristics of the two materials explains why swimmers are leaner, less muscular, and more efficient at propelling themselves through the water.

Mochizuki and Kimura determined in 1952 that the ultimate tensile strength of the panniculus adiposus of pigs was 8 ± 0.4 g/mm² with an

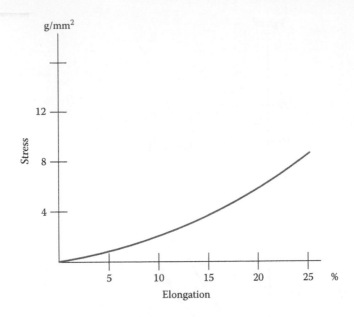

Figure 13.3 Stress–strain in tension of fat.

ultimate elongation of 26% ± 0.7%. For the expansive properties, the ultimate expansive strength was determined to be 0.71 ± 0.023 kg/cm²/mm. The
ultimate expansion for a 7-mm diameter was 0.046 ± 0.0015 mL. Figure 13.3
shows the stress–strain curve in tension and Figure 13.4 shows the pressure–
expansion curve for pig adipose tissue. Adipose tissue is the layer under the
skin and above the muscles and provides reserves during lean times as well
as providing a certain amount of protection from injury. The presence of
adipose tissue should always be considered when making biomechanical calculations as it effectively behaves in a cushioning manner.

Man-Made Protective Structures

We now turn our attention to those devices that have been designed to protect humans from injury as a result of normal activities. As we mentioned,
we will not discuss the safety systems of automobiles such as seat belt, airbags, and antilock brakes. We will not discuss rollover and fall protective
structures for machinery as they actually do not fit into the premise of the
book. Machine guarding and other protective structures designed to prevent
or mitigate injury are also not a part of this discussion. The protective structures listed earlier are more general and designed with engineering safety in
mind rather than personal protective safety.

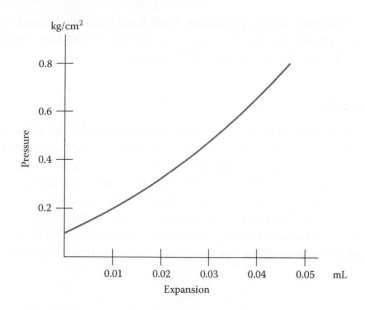

Figure 13.4 Pressure–expansion of adipose tissue.

The term used for the safety of personnel is referred to as personal protective equipment (PPE). The main design function of this type of equipment is to protect workers or individuals involved in a multitude of activities, including the work environment and sport or leisure activities. Again, we are not concerned with equipment built into a device to protect the individuals such as saw blade guards or sheaths on sharp objects. To reiterate, we are only looking at equipment worn by an individual involved in a particular activity.

The equipment that comes to mind in this category is designed to protect individuals from injury or illness resulting from contact with physical, chemical, electrical, mechanical, or radiological hazards in the workplace, in physical or sport activities, or in leisure. This type of equipment includes hard hats, helmets, face shields, safety goggles or glasses, safety shoes, safety garments, including gloves, shoes, vests, ear protection, and respiration protection.

With work-related activities, the employer has the responsibility to provide worker protection. These work-related activities fall under the purview of the Occupational Safety and Health Administration known as OSHA. OSHA's primary responsibility for protective equipment is outlined in the standards in Title 29 of the Code of Federal Regulations (CFR), part 1910 Subpart I. The OSHA Standards are also found in most state regulations and industry standards. Please refer to the applicable standards in Chapter 15.

Personal protective equipment falls under the following categories: protection from head injuries, protection from leg and foot injuries, protection from face and eye injuries, protection from hand injuries, protection from body injury, protection from hearing loss, and protection from respiratory conditions. We briefly outline each of these categories.

Head Injuries

As outlined throughout the book, head injuries may be the most critical. Hard hats and helmets afford a certain amount of protection from head impact, penetration injuries, falling objects, or direct contact with dangerous environments. Not only does OSHA mandate head protection for most work-related environs, but also most sports have recognized the importance of head protection. A review of the applicable standards in Chapter 15 reveals the importance of head protection as outlined in ASTM Standards. Head protection is available for many and varied activities from bicycling to skiing.

Leg and Foot Injuries

Football players are equipped with thigh and knee pads, baseball catchers are equipped with shin, and groin guards as well as chest and face protection. Other sports have a multitude of protective equipment for their legs and feet. Work environments require safety shoes, foot guards, leggings, and leather or rayon protective equipment. These safety devices are meant to protect workers from falling, sharp, or rolling objects. They protect from the dangers of wet and slippery surfaces, hot objects, and electrical hazards. Snow skiing equipment includes legging bindings for the attachment to skis so that when the skis become dislodged in a fall, the bindings give way before ankle and knee injuries occur.

Face and Eye Injuries

Hockey players are now equipped with helmets that have shields, especially the goalies. In addition, they have guards across the face to protect from the dangers of a flying puck. Workers may be exposed to flying fragments, sparks, radiation sand, dirt, and mists so that protection from these hazards is required in most industries In recent years, the dangers of solar radiation have been recognized so that workers or sport enthusiasts and athletes now wear sun glasses. Exposure to significant glare from the sun has been associated with cataract growth, especially at advanced ages for individuals. Protective lotions are also available for the skin relative to exposure to the harmful rays of the sun.

Hand and Arm Injuries

Harmful substances are known to permeate the skin, especially through the cuticles. These substances can be carried through the blood stream where they are filtered in the liver. Long exposure to these chemicals is deleterious to health. These maladies are especially worrisome if the hands or arms have cuts, lacerations, and abrasions. Many workers' activities require gloves that reach the elbow in order to protect the forearm. Welders, in particular, require these types of gloves. A variety of glove types are available for particular exposure depending on the application. Boxers and martial artists employ gloves in order to keep from breaking their hands.

Body Injuries

Some of the most common materials used for whole body protection include plastics, leather, rubber, wool, and cotton. These materials protect workers from the damages posed by contact with heat, radiation, hot surfaces, liquids, body fluids, and waste. Scuba divers utilize wet or dry suits to protect from heat loss in the water. Exposure even to warm water temperatures can lead to hypothermia. The human body operates at about 98°F. Extended exposure at 90°F will lower body temperature and lead to death through hypothermia. Body temperature to maintain life must be above 95°F. At the other end of the scale temperatures above 104°F for adults is considered a medical emergency and must be immediately dealt with and is known as hyperthermia.

Law enforcement officials now are almost universally required to wear bullet proof vests. These devices often made of Kevlar are designed to reduce the impact of high-velocity projectiles such as bullets. Kevlar is also used for other types of body protection including gloves and other clothing accessories.

Hearing Loss

Exposure to high noise levels can cause hearing loss in the work environment and in leisure activities such as trap shooting. Environmental noise regulations limit outdoor noise levels below 60–65 dB(A) and occupational levels are restricted to 85–90 dB(A) for 40 h of exposure. Hearing loss is known as presbycusis and occurs naturally with age but is generally exacerbated by noise pollution. Mechanically, hearing loss occurs because of trauma to the stereocilia of the cochlea. There are a variety of ear protection devices including plugs and earmuffs.

Respiratory Injuries

Breathing air contaminated with dusts, fogs, gases, smoke, and vapors can be quite harmful. The breathing mechanism of humans affords a certain amount of protection with nasal hairs and fluids in the bronchial passages. These naturally occurring protective structures have not evolved to keep the insults of our modern world on our breathing apparatus. Respiratory protection is a must in many industrial environments and is also required when people undertake simple tasks such as painting with a spray device.

Examples of Analysis 14

In this chapter, we dedicate our efforts to some of the most common types of injuries that are encountered in biomechanics. The basic calculations are quite simple. Care must be taken to appropriately size the affected area. Care must also be taken to make sure that the units used are correct since the literature has both MKS and English units. Not only must the proper units be used but also the unit analysis should always be carried out. An example of improper use of equations succinctly exemplifies a common mistake. For this example, consider the simple calculation of energy. The total mechanical energy of a system is the sum of the potential energy and the kinetic energy, or

$$E_t = E_p + E_k$$

where the potential energy is $E_p = mgy$ whose units are ft lb in the English system. The kinetic energy is $E_k = (1/2)mv^2$ whose units are also ft lb. So the units of the total mechanical energy must also be in ft lb. The equation would be written correctly as

$$E_t = mgy + \frac{1}{2}mv^2.$$

The unit analysis in the English system is

$$\text{ft lb} = \frac{\text{lb}}{\text{ft/s}^2}\left(\frac{\text{ft}}{\text{s}^2}\right)(\text{ft}) + \frac{\text{lb}}{\text{ft/s}^2}\left(\frac{\text{ft}^2}{\text{s}^2}\right).$$

Let us assume that a mistake is made in the equation for potential energy, and it is written as momentum, which is the product of mass and velocity or mv. What are the units of mv? The units are

$$\frac{\text{lb}}{\text{ft/s}^2}\left(\frac{\text{ft}}{\text{s}}\right) = \text{lb s}.$$

It is now obvious that foot-pounds are not the same as pound-seconds. The similarity of the equation for potential energy to the momentum equation would explain such a simple mistake. The incorrect equation would be

$$E_t = mv + \frac{1}{2}mv^2$$

This equation is like adding apples to oranges which gives fruit salad and not apples or oranges. This example is from an actual book where, with today's technology, self-publishing is common and is not carefully scrutinized. So, be careful when making calculations because it is easy to make mistakes.

Anterior Cruciate Ligaments

A simplified diagram of the knee showing the location of the cruciate ligaments, the collateral ligaments, the femur, tibia, fibula, the menisci, and the cartilage that covers the end of the bones is found in Chapter 4. With respect to the labeling of these structures, it is important to note the following: anterior means to the front, posterior to the rear, lateral to the outside, and medial to the inside. The anterior cruciate ligament (ACL) prevents anterior and rotational motion of the tibia relative to the femur. The most common causes of ACL tears include impacts to the side of the knee, an overextended knee joint, and quick stops. The majority of the injuries are sports related and are more prevalent in women than in men as determined from the Methodist Sports Medicine Center. This fact is due to the less robust nature of the knee structures in women. Although the ACL can be injured by forces in one direction, these injuries usually occur from a combination of movements. The movements generally include flexion and rotation of the knee. These injuries are caused by the bones of the knee twisting in opposite directions. Such motions as quick stops while moving, change in direction, impact from sports, while running, landing, or lunging are the most common form of injury according to Spindler and Kopf.

The average value of the volume of the ACL varies from 1.9 to 2.2 mm³. The width of the ligament varies between 7 and 9 mm. Consequently, the average cross-sectional area is approximately, $A = 0.25$ mm². Yamada indicates that the tensile properties of elastic ligaments in animals vary between 0.16 and 0.32 kg/mm². Based on these dimensions a computation can be made to determine the forces and stresses that may produce an ACL injury. As an example, let us assume a collision of 10 mph or approximately 14.66 ft/s. The thickness of the area of the ligament is based on the ratio of the area divided by the width, or 0.25 mm² divided by 8 mm or approximately 0.03 mm.

Then, the acceleration is given for a penetration to the ligament of approximately 1 in. as

$$a = \frac{v^2}{2t} = \frac{(14.66)^2}{0.166} = 1300 \text{ ft/s.}$$

The mass of the leg for a 200 lb individual is about one-eighth of the total weight, or

$$m = \frac{25 \text{ lb}}{32.2 \text{ ft/s}^2} = 0.77 \text{ slugs.}$$

The force per ligament is

$$F = ma \cong 250 \text{lb} = 113 \text{ kg.}$$

The tensile stress is the force divided by the area of the knee or

$$\sigma = \frac{113 \text{kg}}{9 \text{ in.}^2} = 0.059 \text{ kg/mm}^2.$$

This value is approximately 18% of that required to injure the ACL. Consequently, such an injury is probably not possible in this scenario.

Minimum Speed Required to Fracture the Tibia and Fibula

The mechanical insult to the lower leg stemming from a pedestrian–vehicle collision is classified as a high-energy injury that involves direct impact, high bending forces, and results in transverse fractures of the tibia and fibula. The minimum speed required to fracture the long bones can be quantified as follows: the bending moment of a long bone is found from

$$M_b = \frac{Y_b I_a}{r}$$

where
   M_b is the bending moment
   Y_b is the ultimate tensile strength of bone
   I_a is the area moment of inertia
   r is the radius of the bone at the breaking point

The leg of a human is approximately 15% of the total weight. The lower leg is approximately one-third of the weight or approximately 5% of the weight of the pedestrian. For example, the medical records indicate that the individual weighed 190 lb, and his height was 6 ft 3 in. The ultimate strength of the fibula is 2.03×10^4 psi and that of the tibia is 2.12×10^4 psi, so that the ultimate strength of both bones would be approximately 4.15×10^4 psi. The bending moment for a combined radius of approximately 0.5 in. can then be computed as

$$Y_b = 4.15 \times 10^4 \text{ lb/in.}^2, \quad r = 0.5 \text{ in.,} \quad I_a = \frac{\pi}{4}(0.5)^4 = 0.049 \text{ in.}^4$$

$$M_b = 4072 \text{ in.lb.}$$

The mass of the leg is

$$m = \frac{w}{g} = \frac{10 \text{ lb}}{32.2 \text{ ft/s}^2} = 0.31 \text{ slugs.}$$

The force required to break the bones that are approximately 12 in. long at mid length is

$$F = \frac{M_b}{6} = \frac{4072}{6} = 678 \text{ lb.}$$

The acceleration is

$$a = \frac{678}{0.31} = 2189 \text{ ft/s}^2.$$

The minimum vehicle speed compressing the leg approximately two inches or 0.166 ft would then be

$$v = \sqrt{(2)(2189)(0.166)} = 26.96 \text{ ft/s} = 19 \text{ mph.}$$

So, if the vehicle is traveling at this speed and strikes the pedestrian leg with the bumper, the injury is most certain.

Hip Injuries

The hip joint consists of the articulation of the femur (thigh bone) with the girdle of the pelvis (coxal bone). The articulating surfaces are those of the head of the femur and the acetabulum. The acetabulum is the cup-shaped socket of the hip bone. The two bones have an improved fit produced by the acetabular labrum. The labrum is a lip-like structure made of fibrous carti-lage attached to the coxal bone. The reinforcements of the hip include the iliofemoral ligament anteriorly and the ischiofemoral ligament posteriorly. Please refer to Figure 14.1a through c.

A chondral lesion is a tear of the cartilage that helps to articulate the joint. In an alleged collision, the claimant reported repair to the left hip as a result of a labral tear and chondral lesion of the acetabulum. This type of injury requires significant forces due to the robust nature of the hip girdle. A predominant cause of this type of injury is luxation (dislocation) of the hip joint. Another cause of this type of injury results from significant wear resulting from heavy lifting or from contact sports. Age also has an effect and may result in osteoarthritis. Generally, high-energy trauma is required to produce these injuries.

In motor vehicle accidents, the direction of the force largely determines the type of injury that occurs. Hip injuries are much more prevalent in side-impact collisions. Frontal collisions require significant speed changes at impact. There are two primary methods that the hip is injured in a frontal

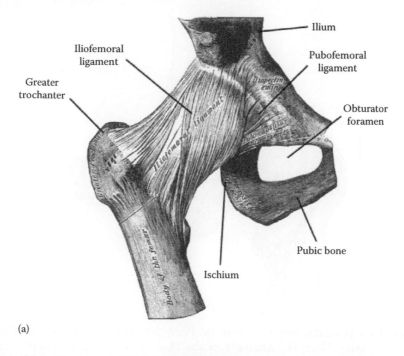

(a)

Figure 14.1 Hip joint: (a) Plate 339 (From "Gray339" by Henry Vandyke Carter, Gray, H., Anatomy of the Human Body, Bartleby.com: Gray's Anatomy, 1918, http://commons.wikimedia.org/wiki/File:Gray342.png#/media/File:Gray342.png, accessed May 12, 2015.)   (*Continued*)

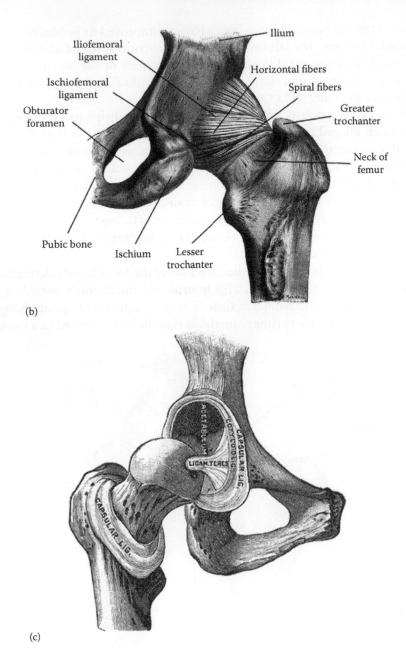

Figure 14.1 (Continued) Hip joint: (b) Plate 340 (From "Gray340" by Henry Vandyke Carter, Gray, H., Anatomy of the Human Body, Bartleby.com: Gray's Anatomy, 1918, http://commons.wikimedia.org/wiki/File:Gray340.png#/media/File:Gray340.png, accessed May 12, 2015.), (c) Plate 342 (From "Gray342" by Henry Vandyke Carter, Gray, H., Anatomy of the Human Body, Bartleby.com: Gray's Anatomy, 1918, http://commons.wikimedia.org/wiki/File:Gray342.png#/media/File:Gray342.png, accessed May 12, 2015.)

collision. One is where the knee strikes the dashboard in unbelted occupants. The other type of collision that involves damage to the hip is when a driver suddenly applies the brake, and the force of the impact is transferred through the leg to the hip as the brake pedal is forcibly applied and the knee is locked. In the accident cited earlier, the collision involved the occupant seated in the rear seat. Consequently, an unbelted passenger in the rear seat would strike the rear of the front seat with his knees. The rear of the front seats is significantly softer and provides much more cushion than the front dashboard or the brake pedal with a locked knee. These types of collisions may involve both luxation and fracture of the hip joint. Dynamic modeling of hip injuries from falls has shown that the forces from such injury mechanisms may produce between 13,000 and 44,000 lb of force. This range of forces is certainly sufficient to cause luxation or fracture of the hip girdle.

Meniscus Tear, Medial, and Lateral

The meniscus distributes body weight across the knee joint. There are two menisci in the knee: the medial (on the inside) and the lateral (on the outside). These structures are C-shaped and have a wedge profile that keeps the rounded ends of the femur from sliding off the flat tibial surface. The meniscus is made of cartilage. Cartilage also covers the ends of the bones allowing for the smooth movement of the bones relative to each other. On the femoral condyle of the knee, the cartilage averages 2.21 mm from studies conducted by Anathasiou et al. Under impact loads, cartilage behaves as a single-phase, incompressible, elastic solid. The aggregate modulus of cartilage varies between 0.5 and 0.9 MPa, and Young's modulus varies between 0.45 and 0.8 MPa. The tensile failure stress of human cartilage varies from 30 to 10 MPa and is a function of age varying from 20 to 80 years. For a 55-year-old, the stress is approximately 17 MPa.

Meniscal injury is often caused by high rates of the application of force according to Whiting. Meniscal injury is either traumatic or degenerative. Traumatic injuries are usually seen in young active individuals. Degenerative tears are chronic injuries generally found in older persons. Damage usually occurs when the meniscus is subjected to a combination of flexion and rotation on extension and rotation during weight bearing and resultant shear between the tibial and femoral condyles. Sports-related meniscal injury is common in soccer, track and field, and skiing. Occupationally, jobs requiring repeated squatting, such as mining, carpet laying, or gardening, are associated with these injuries. In a motor vehicle collision, the mechanism for posterior cruciate ligament injury occurs when the knee impacts the dashboard. This action forces the tibia posteriorly relative to the femur For example, the lower leg of an individual weighing 200 lb weighs approximately 25 lb. The thickness of the meniscus is approximately 2.21 mm or 0.00725 ft. There are two menisci. If the vehicle experiences a speed change of 5 mph or 7.33 ft/s², the acceleration on the meniscus is

$$a = \frac{v^2}{2x} = \frac{(7.33)^2}{(2)(0.0015)} = 1790 \, \text{ft/s}^2.$$

The mass of the lower leg pushing against the meniscus is

$$m = \frac{25 \, \text{lb}}{32.2 \, \text{ft/s}^2} = 0.77 \, \text{slugs}.$$

The force is

$$F = ma = (0.77)(1790) = 1389 \, \text{lb}.$$

The shearing force distributed over both meniscuses is then 1389/2 = 694 lbs. The area of the meniscus is approximately 400 mm². The tensile stress on the meniscus is then

$$\frac{694 \, \text{lb}}{400 \, \text{mm}^2} = \frac{3088 \, \text{N}}{400 \, \text{mm}^2} = 7.7 \, \text{MPa}.$$

Since the tensile failure stress is approximately 17 MPa, the force is approximately 45% of that required for failure.

Rotator Cuff Injuries

In many rear-end collisions, shoulder injuries are claimed. The shoulder contains two bones: the scapula and the clavicle. The clavicle attaches medially to the sternoclavicular joint and laterally to the acromion process. The humerus articulates with the scapula at the glenohumeral joint commonly known as the shoulder joint. Please refer to the diagrams of the shoulder in Chapter 4, which reveal the bones and the muscle attachments.

The most important muscles of the rotator cuff are the subscapularis, supraspinatus, infraspinatus, and teres minor muscles. The muscles attach to the bony structures via tendons. These muscles stabilize the joint by forming a cuff around the humeral head. Several significant injuries to the shoulder occur, including rotator cuff tears. In this case, the claimant was in the front car that was rear ended. The reconstruction revealed that the Lexus GS 350 driven by the claimant was rear-ended by a GMC Savana. The reconstruction based on the damage profiles indicated that the Lexus underwent a speed change of 7.2 mph and the GMC underwent a speed change of 5.1 mph. The medical claim of his injuries was as follows: cervical sprain and strain and partial rotator cuff tear. The average and ultimate tensile stress of tendons and ligaments varies between 50 and 100 MPa or between 14,000 and 7,250 psi. Tendon cross sections vary between 125 and 150 cm² or 19.37 and 23.25 in.² The biomechanical calculations were carried out with his weight varying between 150 and 250 lb with a central value of 200 lb because his actual weight was not known. The speed change was varied from 7 to 11 mph to consider a worst case. The tensile stresses on his shoulder tendons then varied between 4% and 13% of that required to produce injury.

In conclusion, it could be stated with a very high degree of biomechanical engineering certainty that the injuries to the shoulder of the claimant were inconsistent with being produced by the collision. Tendons and ligaments are very resilient structures that can withstand extreme forces. The forces produced by the collisions in this case were very minor in comparison.

Since the alleged injury to the shoulder was the most significant, the biomechanical calculations were not carried out for the cervical sprain and strain that were being alleged. Cervical spinal injuries require a speed change in excess of 12 mph to produce any scientifically proven injury. The previous example on meniscus tears shows the procedure followed for this case. The calculations are left as an exercise for the reader.

Shoulder Injuries in General

There are several main types of shoulder injury. These are acromioclavicular (AC) sprain, rotator cuff pathologies, glenohumeral instability and dislocation, bicep tendinitis, impingement syndrome, and labral pathologies. The more common types are AC sprain and rotator cuff (RC) injuries. AC sprains result from forces that tend to displace the scapular acromion process from the distal end of the clavicle. Commonly, this type of injury is referred to as separated shoulder. A separated shoulder is not the same as a shoulder dislocation, which is more serious. AC injuries may result from direct or indirect forces. The most common cause of AC injury is produced by a direct force applied to the point of the shoulder with the arm in an adducted (toward the median axis) position. Such injuries are produced in falls where the shoulder impacts the floor. Indirect forces that produce AC injuries occur in falls where the outstretched arm is laterally extended and the hand strikes the floor.

Rotator cuff injuries are a common complaint of persons who utilize overhead movements. Glenohumeral impingement and rotator cuff lesions are the most common shoulder conditions because of their morphology. Impingement pathologies fall into two categories: for persons below and above

35 years of age. Younger individuals are susceptible to injury resulting from sports- or work-related activities. Older individuals are more likely to suffer from the effects of degenerative processes that lead to bone spur formation, capsular thinning, decreased tissue perfusion, and atrophy of the muscles. The muscle action of the shoulder joint is designed to keep the reaction force in line with the glenoid. The joint becomes unstable as the line of action moves away from the geometric center of the glenoid. Individuals with lax shoulder muscles may experience glenohumeral dislocation resulting from minimal forces. The vast majority of dislocations occur anteriorly. Anterior luxation occurs most often from indirect forces when the axial loads are applied to the abducted, extended, and internally rotated arm. Anterior dislocation may also occur from direct forces applied to the posterior aspect of the humerus. Posterior glenohumeral dislocation (luxation) may occur from a direct force to the anterior aspect of the shoulder or from an indirect force through the arm if extended, flexed, adducted, or in an internally rotated position.

The shoulder contains two bones, the scapula and the clavicle. The clavicle attaches medially to the sternoclavicular joint and laterally to the acromion process. The humerus articulates with the scapula at the glenohumeral joint commonly known as the shoulder joint. Figure 14.2 shows the shoulder joints and Figure 14.3 shows the muscle attachments.

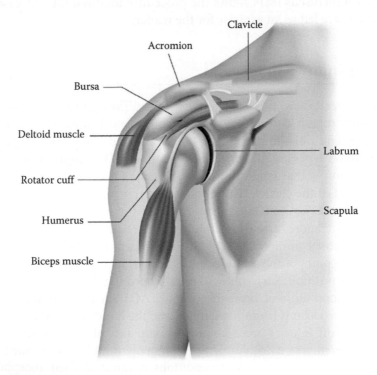

Figure 14.2 Shoulder joint.

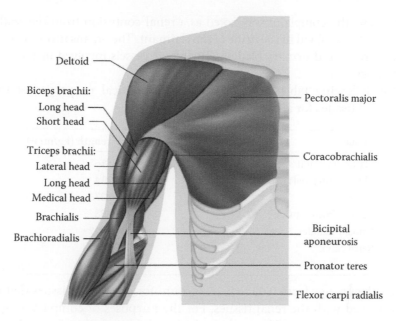

Figure 14.3 Shoulder muscles.

The most important muscles of the rotator cuff are the subscapularis, supraspinatus, infraspinatus, and teres minor. The muscles attach to the bony structures via tendons. These muscles stabilize the joint by forming a cuff around the humeral head. Several significant injuries to the shoulder occur, including labral pathologies. The glenohumeral labrum is a fibrocartilage rim that encircles the articular surface of the scapular glenoid fossa. Labral injuries may be chronic or acute and are caused by a variety of mechanisms, including compressive falls, traction (tension) from lifting, throwing, overhead movements, and dislocation.

The average and ultimate tensile stress of tendons and ligaments varies between 50 and 100 MPa or between 14,000 and 7,250 psi. Tendon cross sections vary between 125 and 150 cm^2 or 19.37 and 23.25 in.^{2} In one case, the reconstruction determined that the accelerations on the claimant's shoulder would vary between 4 and 8 g's. For her body weight of approximately 150 lb, the forces exerted on her shoulders varied between 260 and 511 lb. The tensile stresses on her shoulder tendons then varied between 13.7 and 22.2 psi. These stresses are between 0.19% and 0.31% of that required to produce injury.

Kidneys, Arteries, and Veins

The mechanical properties of urogenital organs and tissues are well documented in the literature. The urogenital organs and tissues include the kidney, ureter, urinary bladder, uterus, vagina, amnion, and umbilical cord.

In this case, the complaint was noted as a renal contusion from the seatbelt, which produced blood in the urine of the claimant. The organs involved are the kidney, ureter, and urinary bladder. The blood vessels involved may be veins and arteries.

The following table summarizes the mechanical properties of those organs and tissues for all age groups.

Tissue	Ultimate Tensile Strength (kg/mm²)
Elastic arterial tissue	0.068–0.162
Muscular arterial tissue	0.09–0.14
Veins	0.12–0.27
Renal fibrous capsule	0.21–0.28
Urinary bladder	0.020–0.042
Renal calyx	0.048–0.107
Ureter	0.042–0.206

These values range from 0.02 to 0.28 kg/mm² for any of the tissues that may be associated with the renal tissues. For the purposes of computations, the value of 0.02 kg/mm² will be used to calculate the possibility of a renal contusion from this accident.

The reconstruction indicated that the speed change of the vehicle was 8.8 mph or 13 ft/s. The normal displacement through which the forces would decelerate to reach these tissues is approximately 4 in. or 0.33 ft. This value is for a 200 lb individual. For heavier individuals, the value increases, thereby decreasing the acceleration. Therefore, the acceleration is

$$a = \frac{v^2}{2x} = \frac{(13)^2}{2(0.33)} = 256 \, ft/s^2 = 8g\text{'s}.$$

The force produced by the seat belt on the abdominal area for a 200 lb individual is distributed on ⅛ of the weight of the person

$$F = ma = \frac{w}{g} a = \frac{(25)(256)}{32.2} = 200 \, lb = 90 \, kg.$$

The area over which this force is distributed by the seat belt is

$$A = (2 \, in.)(16 \, in.) = 32 \, in.^2 = 20,645 \, mm^2.$$

The tensile stress produced on the organs is then

$$\frac{F}{A} = \frac{90 \, kg}{20,645 \, mm^2} = 0.0044 \, kg/mm^2$$

which is approximately 22% of the ultimate tensile stress of the urinary bladder tissue. This injury is probably not possible as alleged.

As a general rule, when carrying out biomechanical calculations, in order to assign a possibility for injury, the stresses need to exceed 75% of the ultimate stress of the particular body section. Keep in mind that you should always do a worst case analysis so that a computation approaching 75% of the ultimate stress may actually be less than the computed value. Keep in mind that normal human activity, in general, does not produce injury.

Teeth

In this biomechanical calculation, it is necessary to determine if a rear-end collision between two vehicles can cause misalignment or injury to teeth if the head moves forward or backward. The claimant was a child with misaligned teeth that required braces. The child was 12 years old with permanent teeth. The parents were hoping that insurance from the minor collision would cover the cost of the orthodontic treatment. It was assumed that the head does not strike any portion of the interior of the vehicle since no such event was reported. However, from the reconstruction of the collision, the speed change of the Honda was determined, based on the crush analysis, as 4.1 mph or 6 ft/s. The speed change of the Dodge was computed to be 3.1 mph or 4.5 ft/s. As a worst case scenario, a speed change of 6 ft/s was assumed although the vehicle in which the child was riding was the Dodge.

In 1974, Hariri et al. determined that the breaking stress varied from 28.62 ± 1.94 and 31.92 ± 2.44 kg/mm^2. Thus, the variability was from 26.68 to 34.36 kg/mm^2. According to Potiket et al. (2004), the breaking strength of ceramic crowns varied between 66.8 and 88.6 kg. The lowest value of 26 kg/mm^2 was used for this analysis. The calculations determine a worst case scenario at the breaking point of a tooth.

The acceleration produced in the earlier-described scenario is a function of the penetration of the tooth into the surrounding tissue of the gums. This penetration distance is at least $\frac{1}{8}$ of an inch or 0.01 ft so that

$$a = \frac{v^2}{2x} = \frac{6^2}{2(0.01)} = 1800\,\text{ft/s}^2 = 56\,g\text{'s}.$$

The human head weighs approximately 5% of the total body weight. If an individual weighs 200 lb, the head would weigh approximately 10 lb. Its mass is

$$m = \frac{w}{g} = \frac{10}{32.2} = 0.31\,\text{slugs}.$$

The force imparted by the acceleration is

$$F = ma = (0.31)(56) = 17.3 \text{ lb} = 7.9 \text{ kg}.$$

The standard tooth size of a human child is approximately 4 mm × 2 mm = 8 mm². Thus, the stress on the tooth would be

$$S = \frac{F}{A} = \frac{7.9}{8} \sim 1.0 \text{ kg/mm}^2.$$

This stress is approximately 4% of that required to break the tooth. Therefore, the forces imparted on a tooth by the scenario described are not sufficient to break much less dislodge or misalign teeth. Note that even if the weight of the head would be 20 lb, a ridiculous value, the stress would be only 8%.

Closed Head Injuries

Consider a person who supposedly suffers severe injury from being hit by a 3 ft section of a 2 × 4 piece of poplar framing lumber that resulted from an altercation. Stalnaker determined that the brain movement would be approximately 3 cm or 0.03 m in such an event. From the discussions of Chapters 4 and 7, we see that severe brain injury occurs when the pressure of the blow exceeds 34 lb/in.² or a corresponding acceleration of 1758 m/s². The velocity of the blow would be

$$v = \sqrt{2ax} = \sqrt{(2)(1758)(0.03)} = 10.3 \text{ m/s} = 22.9 \text{ mph} = 35.6 \text{ ft/s}.$$

A 3ft section of poplar would weigh approximately 2 lb. So the energy of the impact would be

$$E = \frac{1}{2} \frac{2}{(32.2)} (35.6)^2 = 35 \text{ ft lb} = 420 \text{ in. lb}.$$

If the blow produced an indentation on the head that was 1.5 × 3.5 in.² in area or 5.25 in.² at an average depth of 5/16 of an inch, the force would be (5/16) (420) = 131 lb. Then, the pressure is

$$\sigma = \frac{F}{A} = \frac{131}{5.25} = 25 \text{ psi}.$$

Since this value is less than the pressure of 34 psi required, the injury is inconsistent with the claim. This stress is 73.5% of the threshold for closed head injury. However, a slight variation in the values of the computation may make the injury possible.

Tibia Plateau and Eminence Fractures

The following example involves a vehicle–pedestrian accident. A sedan was being operated at low speeds when a pedestrian walked out into traffic with the left side of the pedestrian's body oriented toward the front of the sedan. Due to various factors, the driver was unable to stop to avoid striking the pedestrian. The front bumper of the sedan strucks the outer side of the pedestrian's left knee. After the impact, the pedestrian fell onto her right shoulder.

The vehicle weight, including its two passengers, was approximately 4200 lb. The injured pedestrian was a 5 ft 2 in. tall, 334 lb, 50-year-old female. The pedestrian endured a lateral fracture and depressed fracture of the eminence of the tibia. Additionally, she endured a right proximal humerus fracture. Using available data sources, determine the required speed of the vehicle to cause an eminence fracture. Assess if the knee injury could have been caused by the pedestrian walking into a stationary vehicle. Additionally, determine if the knee and shoulder injuries could have been caused by a fall.

Solution

Consider an exploded view of the knee, where the condyles of the femur are situated atop the tibia plateau (Figure 14.4).

The tibia plateau includes a dual ridge that is oriented from front to back. This ridge, or eminence, fits between the condyles of the femur, which prevents sideways sliding.

From Yamada, the ultimate tensile strength of compact bone is given to be 12.5 kg/mm², or roughly 27.6 lb/mm². For the 50–59-year age group, this strength is reduced to 9.5 kg/mm², or roughly 20.9 lb/mm². Damask states that the area of the eminence fractured by a sideways force is approximately 350 mm². The static force required to fracture the eminence of the tibia plateau is thus

$$F_{fracture} = \sigma A = (27.6 \text{ lb/mm}^2)(350 \text{ mm}^2) = 9660 \text{ lb}$$

$$F_{fracture} = \sigma A = (20.9 \text{ lb/mm}^2)(350 \text{ mm}^2) = 7315 \text{ lb} [50\text{--}59 \text{ year}]$$

Consider a lateral impact from a vehicle into the knee. The height of the knee is consistent with the elevation of the sedan's front bumper. Here, we assume that

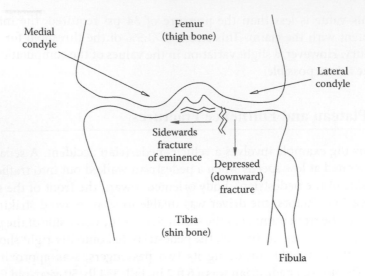

Figure 14.4 Tibia plateau.

there is no sliding of the foot at the road surface during the impact phase. The force of impact is equal to the mass "m" of the object times the acceleration "a."

$$F_{impact} = ma = \frac{W}{g}a$$

The acceleration is equal to the change in velocity "v" squared divided by two times the displacement "x."

$$a = \frac{v^2}{2x}$$

These equations can be rewritten to relate speed change with the force of impact.

$$v = \sqrt{\frac{2Fxg}{W}}$$

For an impact into a rigid surface, which would limit any displacement to compression of the knee and other biological structures, the displacement can be limited to ½ in. By limiting the displacement distance, a minimum impact speed can be determined.

$$v = \sqrt{\frac{2(9660\,\text{lb})(0.04\,\text{ft})(32.2\,\text{ft/s}^2)}{4200\,\text{lb}}} = 2.4\,\text{ft/s} = 1.7\,\text{mph}$$

For the 50–59-year age group, the required impact force to fracture the tibia eminence is 7315 lb. Thus, the impact speed would be reduced to 1.4 mph.

However, modern vehicles have bumpers designed to compress under loading. For a 2 in. compression of the bumper, the displacement distance is increased to about 0.21 ft, while a 3 in. bumper compression is associated with a 0.29 ft displacement. For the maximum impact force of 9660 lb, the required speed to fracture the eminence is 3.8–4.5 mph assuming bumper compression. For the 50–59-year age group, this speed is reduced to 3.3–3.9 mph for the 2–3 in. bumper compression.

Therefore, an impact of 4–5 mph is sufficient to produce a sideward fracture of the tibia eminence. For the case of the pedestrian walking into the vehicle, much higher speeds would be required. Consider a required impact force of 7315 lb and displacement of 2.5 in. of the bumper and ½ in. for the knee. In order to fracture the eminence by walking into a stopped vehicle, the required traveling speed of the pedestrian is as follows:

$$v = \sqrt{\frac{2(7315\,\text{lb})(0.25\,\text{ft})(32.2\,\text{ft/s}^2)}{334\,\text{lb}}} = 18.8\,\text{ft/s} = 12.8\,\text{mph}$$

This traveling speed is highly unlikely given the weight and age of the pedestrian. Thus, it can be stated with confidence that the fracture of the tibia eminence was caused by the sedan striking the pedestrian, rather than the pedestrian walking into a stopped vehicle. However, it is also necessary to examine whether or not the injuries could have been caused by falling.

Consider Figure 14.5, which represents a simplified model of a falling object. The object has length "*l*" with a radial distance "*r*" measured from the point of tipping.

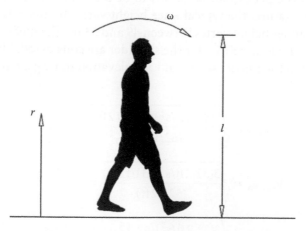

Figure 14.5 Falling object.

The object falls at a rotational velocity "ω." The mass moment of inertia "I" is defined as follows:

$$I = \frac{ml^3}{3}$$

Conservation of energy principles state that the potential energy of falling from a height "h" is equal to the kinetic energy produced by the rotational velocity:

$$PE = KE \Rightarrow mgh = \frac{1}{2}I\omega^2$$

Here, the maximum fall height "h" is equal to the object length "l." The rotational velocity "ω" is equal to the ratio of linear velocity "v" to the radial measurement "r." Manipulating these equations yields the following relationship:

$$v = \sqrt{\frac{3gr^2}{l}}$$

For the case in which r = l, the maximum fall speed is therefore:

$$v = \sqrt{3gl}$$

The height of the pedestrian is known to be 5 ft 2 in., or 62 in. From data tables for this stature, the typical knee height varies between 17 and 18 in., while the shoulder height varies between 49 and 53 in. The median heights of 17.5 in. for the knee and 51 in. for the shoulder are considered. The following equations detail the impact speeds at each elevation during a fall:

$$v_{knee} = \sqrt{\frac{3(32.2\,\text{ft/s}^2)(17.5/12\,\text{ft})}{(62/12\,\text{ft})}} = 6.3\,\text{ft/s}$$

$$v_{shoulder} = \sqrt{\frac{3(32.2\,\text{ft/s}^2)(51/12\,\text{ft})}{(62/12\,\text{ft})}} = 18.4\,\text{ft/s}$$

$$v_{max} = \sqrt{3(32.2\,\text{ft/s}^2)(62/12\,\text{ft})} = 22.3\,\text{ft/s}$$

For an impact of the knee into the ground, the displacement distance may be limited to 1 in. Thus, the acceleration associated with this impact is

$$a = \frac{v^2}{2x} = \frac{(6.3\,\text{ft/s})^2}{2(1/12\,\text{ft})} = 238\,\text{ft/s}^2 = 7.4\,g\text{'s}.$$

Assuming that the full weight of the pedestrian acts on the knee, the force of impact is given by

$$F = ma = \left(\frac{334\,\text{lb}}{32.2\,\text{ft/s}^2}\right)(238\,\text{ft/s}^2) = 2470\,\text{lb}.$$

As documented, the required impact load to fracture the tibia eminence is between 7,315 and 9,660 lbs. Thus, a fall onto the knee is not sufficient to cause this injury.

Conversely, the humerus fracture may be attributed to the fall, given greater fall speeds for the upper portions of the body. From Yamada, the modulus of elasticity for the humerus is 1750 kg/mm². The inner and outer radius of this bone, as taken from data sources, is 3.7 and 11.2 mm, respectively. The cross-sectional area of the bone is given by the following:

$$A = \pi\left(r_o^2 - r_i^2\right) = 351\,\text{mm}^2$$

For a tensile strength of 12.5 kg/mm², the required fracture load is 9674 lb. For the 50–59-year age group, wherein the tensile strength is reduced to 9.5 kg/mm², the required fracture load is 7353 lb.

From before, we know that the falling speed at the elevation of the shoulder is about 18.4 ft/s. For a 1-in. displacement, the acceleration is given by the following:

$$a = \frac{v^2}{2x} = \frac{(18.4\,\text{ft/s})^2}{2(1.0/12\,\text{ft})} = 2031\,\text{ft/s}^2 = 63\,g\text{'s}$$

For a worst-case scenario, the entire weight the pedestrian is assumed to be falling onto the shoulder. Therefore, the impact force required to fracture the humerus is most near:

$$F = ma = \left(\frac{334\,\text{lb}}{32.2\,\text{ft/s}^2}\right)(2,031\,\text{ft/s}^2) = 21,070\,\text{lb}$$

This force can be compared to the limiting loads of 7353–9674 lb. This analysis is simplified by the assumption that all the weight of the pedestrian acts on the shoulder during the fall. Actual falls may be more complicated and entail initial impacts about the knee and hip areas that reduce the effective load acting on the shoulder.

Cervical Injuries: A Comparison

In this example, two methods of analysis are compared for the potential for injury of the cervical spine. This type of injury is common in rear-end collisions where the struck vehicle passengers may experience "whiplash." The scientific literature on the speed change required to produce spinal soft tissue injury places the onset at about 12 mph. At 17 mph the possibility for injury is essentially 100%. Damask has suggested a model to determine the possibility for spinal injury as follows: the differential distance between the lower spine and the rest of the spine is approximately 0.2 ft the acceleration is then

$$a = \frac{v^2}{2x}.$$

The mass of an individual for this model is determined from the weight (w) of the individual divided by two to consider only the torso weight and then multiplied by 5/6 of the torso weight. Then the mass is

$$m = \frac{w}{2}\left(\frac{5}{6}\right) = \frac{5w}{12}.$$

The force is then

$$F = ma.$$

The shearing force is then assumed to be evenly distributed over 15 discs, 7 cervical, and 8 thoracic. The force in kg is then

$$F_k = \frac{F}{15(2.2)}.$$

The tensile stress on a disc is then divided by the cross section of a disc, which is about 660 mm².

$$\sigma = \frac{F_k}{660}.$$

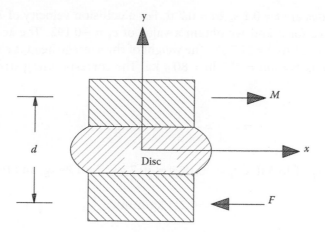

Figure 14.6 Couple model.

For a 200 lb individual at a speed change of 13.6 mph or 20 ft/s the accelera-tion is 1000 ft/s². The mass is 2.58, the force per disc is 78.4 kg, and the stress is 0.118 kg/mm². If we compute the stress at 18 mph or 26.3 ft/s, the accel-eration is 1729 ft/s², the force is 135 kg, and the stress is 0.205 kg/mm². This second value of the stress calculated is slightly greater than the lowest value of ultimate stress of 0.20 ± 0.03 kg/mm² listed in Chapter 8.

An alternate method is as follows. Consider Figure 14.6, which repre-sents two vertebrae separated by a disc. The analysis is that of a couple that produces a moment as follows.

It can be shown that the velocity of the impact and the acceleration that is produced are determined from the following equations:

$$v = \frac{-1}{((t/2d) + c_1)}$$

$$a = \frac{1}{2d((t/2d) + c_1)^2}$$

where
  v is the velocity (ft/s)
  a is the acceleration (ft/s²)
  t is the acceleration pulse width (s)
  d is the distance between vertebrae (ft)
  c_1 is the integration constant

Typical values are $t = 0.1$ s, $2d = 0.7$ ft. For a collision velocity of 20 ft/s, we need to solve for c_1 and we obtain a value of $c_1 = -0.192$. The acceleration becomes $a = 571$ ft/s^2 = 17.7 g's. The weight of the average head is about 10 lb, so the force is $F = ma = 177$ lb = 80.4 kg. The corresponding stress on the disc is

$$\sigma = \frac{F}{A} = \frac{80.4}{660} = 0.121 \, \text{kg/mm}^2.$$

For a velocity of 26.3 ft/s, $c_1 = -0.18$, $a = 30.7$ g's, $F = 139$ kg, and the stress is

$$\sigma = 0.211 \, \text{kg/mm}^2.$$

Note that the values obtained from the couple model are slightly higher than those from the Damask analysis and are not dependent on the weight of the individual. They are dependent on the weight of the head of the individual. Recall that the minimum stress in tension for a disc is $\sigma = 0.20 \pm 0.02$. Both methods indicate that injury to the disc would most probably occur at an impact velocity of 18 mph = 26.3 ft/s but not likely at an impact velocity of 13.6 mph = 20 ft/s.

These computations are in agreement with the literature that indicates the threshold for cervical injury is at about 12–13 mph and significant likelihood of injury at 17–18 mph.

Federal and Other Standards

15

There are a number of standards that are useful when performing biomechanical calculations. There are basically two types of standards. These are federal standards and industry standards. A brief explanation of how standards are applied in the United States is necessary along with the distinction between the two types of standards.

Federal standards are referred to as the Code of Federal Regulations (CFR). These standards cover a very wide range of topics for virtually all parts of the government. These standards are inclusive for all the states in the union and the territories of the United States. The range of topics applicable to biomechanical analysis includes safety standards for materials, equipment, labor, and transportation. A vital aspect of CFR is that it also incorporates other standards by reference. In the various CFR titles, chapters, and sections, many other standards are incorporated in this manner. Make no mistake, the CFR is the law of the land.

Most of the standards available in forensic biomechanical engineering come from the industry as well as from the federal government. The industry standards are from technical and professional societies and from testing facilities. One professional society is the National Fire Protection Association (NFPA). NFPA promulgates a variety of standards basically related to life and safety. Life and safety are, if you recall, the primary objective in engineering. The principal underlying NFPA Code is NFPA 101: Life Safety Code. This code is the umbrella code for all other industrial codes, which are referred to as ANSI Codes. ANSI stands for American National Standards Institute. ANSI includes standards from the American Institute of Electrical Engineers, American Society of Mechanical Engineers, American Society of Civil Engineers, American Institute of Mining and Metallurgical Engineers, American Society of Testing and Materials (ASTM), and the U.S. Departments of the Navy, War, and Commerce. All of the standards promulgated by these various technical societies are considered ANSI standards that by reference tie into NFPA 101. Essentially, all these standards are also part of the law of the land.

Federal Standards

Applicable standards from CFR are the following:

Title 1 General Provisions, *Chapter II*, Part 51 Incorporation by Reference

Title 16 Commercial Practices, *Subchapter B*, Part 1201 Safety Standards for Architectural Glazing Materials, Part 1203 Safety Standards for Bicycle Helmets, Part 1205 Safety Standard for Walk-Behind Power Lawn Mowers, Part 1207 Safety Standards for Swimming Pool Slides, Parts 1213–1226 Safety Standards for Children Appurtenances, Part 1420 Requirements for All Terrain Vehicles, Part 1512 Requirements for Bicycles

Title 23 Highways, *Chapter II*, *Subchapter B*, Part 1215 Use of Safety Belts

Title 29 Labor, *Chapter XVII*, Part 1910 Occupational Health and Safety Standards

Title 30 Mineral Resources, *Chapter I*, *Subchapter O*, Part 70 Mandatory Health Standards, Underground Coal Mines; Part 71 Mandatory Health Standards, Surface Coal Mines and Surface Work Areas of Underground Coal Mines; Part 72 Health Standards for Coal Mines; Part 74 Coal Mine Dust Sampling Devices; Part 75 Mandatory Safety Standards, Underground Coal Mines; Part 77 Mandatory Safety Standards, Surface Coal Mines and Surface Work Areas of Underground Coal Mines

Title 49, Transportation, *Chapter II*, Parts 214 Railroad Workplace Safety, Part 215 Railroad Freight Car Safety Standards; *Chapter III*, Part 393 Parts and Accessories Necessary for Safe Occupation; *Chapter V*, Part 571 Federal Motor Vehicle Safety Standards

Industry Standards

SAE Standards

SAE J98 Personnel Protection for General-Purpose Industrial Machines

SAE J231 Minimum Performance Criteria for Falling Object Protective Structures

SAE J386 Operator Restraint Systems for Off-Road Work Machines

SAE J1040 Performance Criteria for Rollover Protective Structures for Construction

SAE J1042 Operator Protection for General-Purpose Industrial Machines

SAE J1388 Personnel Protection—Steer Skid Loaders

SAE J2426 Occupant Restraint System Evaluation—Lateral Rollover System-Level Heavy Trucks

ASTM Standards

F355-10a Test Method for Impact Attenuation of Playing Surface Systems and Materials

F381-13 Safety Specification for Components, Assembly, Use, and Labeling of Consumer Trampolines

F429-10 Test Method for Shock-Attenuation Characteristics of Protective Headgear for Football

F513-12 Specification for Eye and Face Protective Equipment for Hockey Players

F609-01 Test Method for Using a Horizontal Pull Slipmeter (HPS)

F717-10 Specification for Football Helmets

F803-11 Specification for Eye Protectors for Selected Sports

F910-04(2010) Specification for Face Guards for Youth Baseball

F1045-07 Performance Specification for Ice Hockey Helmets

F1163-13 Specification for Protective Headgear Used in Horse Sports and Horseback Riding

F1447-12 Specification for Helmets Used in Recreational Bicycling of Roller Skating

F1492-08 Specification for Helmets Used in Skateboarding and Trick Roller Skating

F1587-12a Specification for Head and Face Protective Equipment for Ice Hockey Goaltenders

F1637-13 Practice for Safe Walking Surfaces

F1694-09 Guide for Composing Walkway Surface Investigation, Evaluation, and Incident Report Forms for Slips, Stumbles, Trips, and Falls

F1937-04(2010) Specification for Body Protectors Used in Horse Sports and Horseback Riding

F1952-10 Specification for Helmets Used for Downhill Mountain Bicycle Racing

F2033-06(2011) Specification for Helmets Used for BMX Cycling

F2040-11 Specification for Helmets Used for Recreational Snow Sports

F2048-00(2011) Practice for Reporting Slip Resistance Test Results

F2232-09 Test Method for Determining the Longitudinal Load Required to Detach High Heels from Footwear

F2397-09 Specification for Protective Headgear Used in Martial Arts

F2400-06(2011) Specification for Helmets Used in Pole Vaulting

F2413-11 Specification for Performance Requirements for Protective (Safety) Toe Cap Footwear

F2530-11 Specification for Protective Headgear with Faceguard Used in Bull Riding

F2681-08(2012) Specification for Body Protectors Used in Equine Racing

F2812-12 Specification for Goggle- and Spectacle-Type Eye Protectors for Selected Motor Sports

F2913-11 Test Method for Measuring the Coefficient of Friction for Evaluation of Slip Performance of Footwear and Test Surfaces/Flooring Using a Whole Shoe Tester

NFPA Standards

NFPA 70E Standard for Electrical Safety in the Workplace

NFPA 101 Life Safety Code

NFPA 301 Code for Safety to Life from Fire on Merchant Vessels

Appendix A: Values of Fundamental Constants

Quantity	Symbol	Value
Acceleration due to gravity	g	9.806 m/s^2
Atomic mass	M_a	1.66×10^{-27} kg
Avogadro's constant	N_0	6.022×10^{23} molecules/mol
Boltzmann's constant	k	1.38×10^{-23} J/K
Conductance quantum	G_o	7.748×10^{-5} S
Electric constant	E_o	8.854×10^{-12} F/m
Electron mass	m_e	9.109×10^{-31} kg
Electron volt	eV	1.602×10^{-19} J
Electron charge	q_e	-1.602×10^{19} C
Faraday constant	F	9.6485×10^4 C/mol
Fine structure constant	α	7.297×10^{-3}
Magnetic constant	μ_0	$4\pi \times 10^{-7}$ N/A^2
Magnetic flux quantum	Φ_0	2.067×10^{-15} Wb
Molar gas constant	R_0	8.314 J/(mol K)
Molar mass constant	M_μ	10^{-3} kg/mol
Newtonian gravitational constant	G	6.673×10^{-11} m^3/(kg s^2)
Planck's constant	h	6.625×10^{-23} J s
Proton mass	m_p	1.672×10^{-27} kg
Speed of light in a vacuum	C_0	2.9979×10^8 m/s
Stefan–Boltzmann constant	σ	5.67×10^{-8} W/(m^2 K^4)
Standard atmosphere	1 atm	1.013×10^5 Pa (N/m^2)
Standard state pressure	ssp	10^5 Pa
Zero temperature	K	-273.15°C

Appendix B:
Conversion Factors

Length
    1 in. = 2.54 cm
    1 ft = 30.48 cm
Area
    1 cm^2 = 0.155 $in.^2$
    1 m^2 = 10.76 ft^2
Volume
    1 ft^3 = 0.0283 m^3
    1 L = 1000 cm^3 = 0.2462 gal
    1 $in.^3$ = 16.39 cm^3 = 4.329×10^{-3} gal
Velocity
    1 ft/s = 0.6818 mph
Acceleration
    1 m/s^2 = 3.281 ft/s^2
Force
    1 lb = 4.448 N = 4.448×10^5 dynes = 0.4535 kg
Mass
    1 g = 6.85×10^{-5} slug = 10^{-3} kg
    1 slug = 32.17 lb
Pressure
    1 atm = 14.7 psi = 1.013×10^6 $dynes/cm^2$ = 29.92 in Mercury at 0°C
    Pa = N/m^2
Energy
    1 J = 10^7 ergs = 0.239 cal = 9.48×10^{-4} Btu = 0.7376 ft lb = 2.778×10^{-4} W h
    1 N m = 0.7376 ft lb
    1 erg = 7.367×10^{-8} ft lb

Explanation of Abbreviations

in. = inches
ft = feet
cm = centimeters
m = meters

L = liters
gal = gallons
mph = miles per hour
lb = pounds
s = seconds
kg = kilograms
N = Newtons
atm = atmospheres
J = Joules
cal = calories
Btu = British thermal units
W = Watts
h = hours
Pa = Pascals
g = gram

Bibliography

Aiello, L. and C. Dean. 1990a. *An Introduction to Human Evolutionary Anatomy.* London, U.K.: Academic Press.

Aiello, L. and M.C. Dean. 1990b. *Human Evolutionary Anatomy.* London, U.K.: Academic Press.

Allen, M.E. et al. 1994. Acceleration perturbations of daily living. A comparison to whiplash. *SPINE* 19(11):1285–1290.

Antmann, V.E. 1971. *Mechanical Stress, Functional Adaptation and the Variation Structure of the Human Femur Diaphysis.* Ergab Anat Entwick, Vol. 44. Berlin, Germany: Springer-Verlag.

Aoji, O. 1959. Metrical studies on the lamellar structure of human long bones. *J. Kyoto Pref. Med. Univ.* 65:941–965.

Athanasion, K.A. et al. 1991. Interspecies comparison of in situ intrinsic mechanical properties of distal femoral cartilage. *J. Orthop. Res.* 9:330–340.

ASTIA (118222) Wright Air Development Center. Wright Patterson Air Force Base, OH.

Barter, J.T. 1957. Estimation of the mass of body segments. WADC Technical Report 57–260 ASTIA (118222). pp. 57–260. Wright Air Development Center, Wright Patterson Air Force Base, OH.

Beer, F. and E.R. Johnston, Jr. 1962. *Vector Mechanics for Engineers: Statics and Dynamics.* New York: McGraw-Hill.

Benedek, G.B. and F.M. Villars. 2000. *Physics with Illustrative Examples from Medicine and Biology,* 2nd edn. New York: Springer-Verlag.

Bloecker, K. et al. 2011. *BMC Musculoskelet. Disord.,* Biomed Central LTD. 12:248.

Bogin, B. 1999. *Patterns of Human Growth,* 2nd edn. Cambridge, U.K.: Cambridge University Press.

Bureau of Labor Statistics. Fatal workplace injuries in 1992. Washington, DC: Bureau of Labor Statistics. Current and revised data.

Burstein, A.H. et al. 1976. Aging of bone tissue: Mechanical properties. *J. Bone Joint Surg.* 58A:81–86.

CFR 49 Code of Federal Regulations. Part 1910. Occupational Safety and Health Standards. Washington, DC.

Chapman, A.E. 2008. *Biomechanical Analysis of Fundamental Human Movements.* Champaign, IL: Human Kinetics.

Cossa, F.M. and M. Faviani. 1999. Attention in closed head injury: A critical review. *Italian J. Neurolog. Sci.* 20(2):145–153.

Craig, M. Jaw loading response of current ATD's. Head Injury Biomechanics. *SAE Int.* 1:37.

Damask, A.C. and J. Damask. 1990. *Injury Causation Analysis: Case Studies and Data Sources.* Charlottesville, VA: The Michie Company.

Dempster, W.T. and G.R.L. Gaughran. 2005. Properties of body segments based on size and weight. *Am. J. Anat.* 120:33–54.

Den Hartog, J.P. 1952. *Advanced Strength of Materials.* Mineola, NY: Dover Publishing Company.

Dorlands Medical Dictionary. 2012. 32nd edn. Atlanta, GA: Elsevier.

Douglas, J.D. and R. Barden. 2003. *Numerical Methods,* 3rd edn. Pacific Grove, CA: Thompson Brooks/Cole.

Drake, R.L., A.W. Vogl, and A.W.M. Mitchell. 2005. *Grays Anatomy.* Atlanta, GA: Elsevier.

Drucker, D.C. 1967. *Introduction to Mechanics of Deformable Solids.* New York: McGraw-Hill.

Duggar, B.C. 1962. The center of gravity of the human body. *Hum. Factors* 4:131–148.

Elert, G. ed. 2006. Size of a human: Body proportions. The Physics Factbook.

Evans, F.G. 1953. Methods of studying the biomechanical significance of bone loss. *Am. J. Phys. Anthropol.* 11:413–434.

FMUSS 105. Federal Motor Vehicle Safety Standard 105. Washington, DC: US Department of Transportation. Current version.

FMUSS 121. Federal Motor Vehicle Safety Standard 121. Washington, DC. US Department of Transportation. Current version.

Frederickson, R. 2007. Influence of impact speed on head and brain injury outcome in vulnerable road user impacts on the car hood. *STAPP Car Crash J.* 51:235.

Fujikawa, K. 1963. The center of gravity in parts of the human body. *Okajimas Fol. Anat. Jap.* 39:117–126.

Guettler, J.H. et al. 2004. Osteochondral defects in the human knee: Influence of defect size on cartilage rim stress and load redistribution to surrounding cartilage. *Am. J. Sport Med.* 32(6):1451–1458.

Hariri, E., I. Gedalea, A. Simkins, and G. Robin. January 1974. Breaking strength of fluoride-treated dentin. *J. Dental Res.* 53:149.

Hazama, H. 1950. Study on the torsional strength of the compact substance of human beings. *J. Kyoto Pref. Med. Univ.* 60:167–184.

Hazama, H. 1956. Study on the torsional strength of the compact substance of human beings. *J. Kyoto Pref. Med. Univ.* 60:167–184.

Higdon, A. et al. 1968. *Mechanics of Materials.* New York: John Wiley & Sons.

Hoffman, J.D. 2001. *Numerical Methods for Engineers and Scientists,* 2nd edn. New York: CRC Press.

Human Tolerance to Impact Conditions as Related to Motor Vehicle Design. April 1980. Warrandale, PA: Society of Automotive Engineers J885. Current version.

Jozasa, L. and P. Kannus. 1997. *Human Tendons, Anatomy, Physiology, and Pathology.* Champaign, IL: Human Kinetics.

Katake, K. 1961a. Studies on the strength of human skeletal muscle. *J. Kyoto Pref. Med. Univ.* 69:463–483.

Katake, K. 1961b. The strength for tension and bursting of human fascia. *J. Kyoto Pref. Med. Univ.* 69:484–488.

Katsura, M. 1959. Study on the strength of human teeth. *J. Kyoto Pref. Med. Univ.* 66:207–219.

Katsura, M. and K. Shono. 1959. Comparison of the mechanical properties of compact bone between proximal and distal bones. *J. Kyoto Pref. Med. Univ.* 66:235–243.

Kazarian, L.E. 1970. Biomechanics of vertebral columns and organ response to seated spinal impact on rhesus monkeys. In *Proceedings of the 14th STAPP Car Crash Conference*, Ann Arbor, MI. Warrendale, PA: Society of Automotive Engineers.

Kimura, H. 1952. Tension test upon the compact substance in the long bones of cattle extremities. *J. Kyoto Pref. Med. Univ.* 51:365–372.

Kimura, T. 1971. Cross-section of human lower limb bones viewed from strength of materials. *J. Anthrop. Soc. Nippon* 79:323–336.

Ko, R. 1953. The tension test upon the compact substance of the long bones of human extremities. *J. Kyoto Pref. Med. Univ.* 53:503–529.

Ko, R. and M. Takigawa. 1953. The tension test upon the costal cartilage of a human body. *J. Kyoto Pref. Med. Univ.* 53:93–96.

Kontulainen, S. et al. 2007. Analyzing cortical bone cross-sectional geometry by peripheral QCT: Comparison with bone histomorphometry. *J. Clin. Densitom.* 10(1):86–92.

Kopf, S. et al. January 2011. Size variability of the human anterior cruciate ligament insertion sites. *Am. J. Sports Med.* 39(1):108–113.

Kubo, K. 1959. Investigation on the development in mechanical strength of various organs. *J. Kyoto Pref. Med. Univ.* 66:433–456.

Kuruma, R. 1956a. On the bending test upon the compact bones. *J. Kyoto Pref. Med. Univ.* 59:3–20.

Lazenby, R.A. 2002. Prediction of cross-sectional geometry from metacarpal radiogrammetry: A validation study. *Am. J. Hum. Biol.* 14:24–80.

Lin, T.W. et al. 2004. Biomechanics of tendon injury and repair. *J. Biomech.* 37(6):865–877.

Lissner, H.R., M. Lebow, and F.G. Encass. 1960. Experimental studies on the relation between acceleration and intercranial pressure changes in mass. *Surg. Gyn. Obstet.* 3:329.

Maffrelli, N. 1999. Rupture of the Achilles tendon. *J. Bone Joint Surg. Am.* 81:1019–1036.

Marchi, D. and S. Tarli. 2004 Cross-sectional geometry of the limb bones of the Hominoidea by hyplanar radiography and moulding techniques. *J. Anthropolog. Sci.* 82:89–102.

Martin, R.B. and P.J. Alkinson. 1977. Age and sex related changes in the structure and strength of the human femoral shaft. *J. Biomech.* 10:233–231.

McClintock, F.A. and A.S. Agon. 1966. *Mechanical Behaviour of Materials.* Reading, PA: Addison-Wesley.

McConnell, W.E. et al. March 1993. Analysis of human test subject kinematic responses to tow velocity rear impacts. SAE Paper 930889. Warrendale, PA: Society of Automotive Engineers.

Mertz, H.J. and L.M. Patrick. 1971. Strength and response of the human neck. SAE Paper 710855. Warrendale, PA: Society of Automotive Engineers.

Mochizuki, T. and H. Kimura. 1952. Tension and bursting tests upon the fat membrane, paniculus adeposus of a pig. *J. Kyoto Pref. Med. Univ.* 51:577–579.

Morgan, J.R. and D.L. Ivey. 1987. Analysis of utility pole impacts. SAE Paper 390607. Warrendale, PA: Society of Automotive Engineers.

Motoshinma, T. 1960. Studies on the strength for bending of human long extremity bones. *J. Kyoto Pref. Med. Univ.* 68:1377–1397.

Murray, E.A. et al. 1994. Acceleration perturbations of daily living. A comparison to whiplash. *SPINE* 19(11):1285–1290.

Muto, T. 1951. Basic study on the compressive test of compact bone. *J. Kyoto Pref. Med. Univ.* 49:567–589.

Nashun, A.M. and R. Smith. 1977. Intercranial pressure dynamics during head impact. In *Proceedings of 21st STAPP Car Crash Conference*, New Orleans, LA. Warrendale, PA: Society of Automotive Engineers.

National Cancer Institute (NIH). SEER training module. http://training.seer.cancer.gov/anatomy/body/terminology.html# directional (accessed May 12, 2015).

Nose, K. 1961. The impact bending strength of human tooth. *J. Kyoto Pref. Med. Univ.* 69:1951–1956.

Oda, M. 1955. The influence of the environment upon the strength of a compact bone. *J. Kyoto Pref. Med. Univ.* 157:1–24.

Okamoto, T. 1955. Mechanical significance of components of bone tissue. *J. Kyoto Pref. Med. Univ.* 58:1004–1006.

O'Neill, M.C. and C.B. Ruff. 2004. Estimating human long bone cross-sectional geometric properties: A comparison of non-invasive methods. *J. Hum. Evol.* 47:221–235.

Oya, H. 1960. Examination of the effects of the condition of materials to the strength test of tissues. *J. Kyoto Pref. Med. Univ.* 67:1337–1364.

Pang, B.S.F. and M. Ying. 2006. Sonographic measurement of Achilles tendons in asymptomatic subjects. *J. Ultrasound Med.* 2:1291–1296.

Pike, J.A. 1990. *Injuries, Anatomy, Biomechanics and Federal Regulations Seminar.* Warrendale, PA. Society of Automotive Engineers.

Piziali, R.L. et al. 1980. Geometric properties of human leg bones. *J. Biomech.* 13:881–885.

Potiket, N., G. Chiche, and I.M. Fruger. November 2004. In vitro Fracture strength of teeth restored with different all-ceramic crown systems. *J. Prosthet. Dent.* 92(5):491–495.

Roberts, D.F. 1978. *Climate and Human Variability*, 2nd edn. Menlo Park, CA: Cummings.

Roy, K.J. 1988. Force, pressure, and motion measurements in the foot: Current concepts. *Clin. Pediatr. Med. Surg.* 5:491–508.

Ruff, C. 2002. Variation in human body size and shape. *Ann. Rev. Anthropol.* 31:211–232.

Ruff, C.B. 2000. Body size, body shape, and long bone strength in modern humans. *J. Hum. Evol.* 38:269–290.

Ruotolo, C., J.E. Fow, and W.M. Notlage. 2004. The supraspinatus footprint: An anatomic study of the supraspinatus insertion. *Arthroscopy.* Pub Med PMID: 15007313. pp. 246–249.

Rybicki, E.F., F.A. Sumonen, and E.B. Weis, Jr. 1972. On the mathematical analysis of stress in the human femur. *J. Biomech.* 5:203–215.

Sears, F. and M. Zemansky. 1964. *University Physics.* Reading, PA: Addison-Wesley.

Sharma, P. and N. Maffulli. 2005. Tendon injury and tendinopathy: Healing and Repair. *J. Bone Joint Surg.—Am.* 87A(1):187–202.

Shepherd, S. 2004. Head trauma. Emedicine.com.

Sonada, T. 1962. Studies on the strength for compression, tension, and torsion of the human vertebral column. *J. Kyoto Pref. Med. Univ.* 71:659–702.

Spector, W.S. ed. 1956. *Handbook of Biological Data.* Philadelphia, PA: W.B. Sanders.

Stalnaker, R.L. et al. 1977. Head impact response. In *Proceedings of the 21st STAPP Car Crash Conference*, New Orleans, LA. Warrendale, PA: Society of Automotive Engineers.

Steidal, R.F. Jr. 1989. *An Introductions to Mechanical Vibrations*, 3rd ed. New York: John Wiley & Sons.

Stone, K.R. et al. May 2007. Meniscal sizing based on gender, height, and weight. *J. Arthrosc. Relat. Surg.* 23(5):503–508.

Szaho, T.J. et al. 1994. Human occupant kinematic response to low speed rear impact. SAE Paper 940532. Warrendale, PA: Society of Automotive Engineers.

Takezono, K., H. Yasuda, and M. Maeda. 1964. The impact snapping strength of human and animal compact bones. *J. Kyoto Pref. Med. Univ.* 73:72–74.

Takigawa, M. 1953. Study upon strength of human and animal tendons. *J. Kyoto Pref. Med. Univ.* 53:915–933.

Tsuda, K. 1957. Studies on the bending test and impulsive bending test on human compact bone. *J. Kyoto Pref. Med. Univ.* 61:1001–1025.

Tsunoda, S. 1950. Study on the hardness test of the bone. *Kyoto: MITT Med. Akad.* 47:103–115.

Ward, C., M. Chan, and A. Nahum. 1980. Intercranial pressure—A brain injury criterion. In *24th STAPP Car Crash Conference*, Detroit, MI. Warrendale, PA: Society of Automotive Engineers.

Wheeless, C.R. 2012. Anatomy of the posterior cruciate ligament. In *Wheeler's Textbook of Orthopaedics.* Durham, NC: Duke University.

Whiting, W.C. and R.F. Zernicke. 2008. *Biomechanics of Musculoskeletal Injury*, 2nd edn., Vol. 6. Champaign, IL: Human Kinetics. pp. 177–178.

Yamada, H. 1941. Die mechanischen eigenschaften der knochen, bedonders beim zugversuch. *Kyoto: MITT Med. Akad.* 33:263–320.

Yamada, H. 1965. Strength of tendon. *Collagen Symp.* 5:117–136.

Yamada, H. 1970. *The Strength of Biological Materials.* Philadelphia, PA: Williams and Wilkins.

Yamada, H. and T. Motoshima. 1960. The directional difference in the strength of compression of the shaft of human extremity bones. *J. Kyoto Pref. Med. Univ.* 68:1398–1404.

Yamaguchi, T. 1960. Study on the strength of human skin. *J. Kyoto Pref. Med. Univ.* 67:347–379.

Yokoo, S. 1952a. Compression test of the cancellated bone. *J. Kyoto Pref. Med. Univ.* 51:273–276.

Yokoo, S. 1952b. The compression test of the costal cartilage of a human body. *J. Kyoto Pref. Med. Univ.* 51:266–272.

Yokoo, S. 1952c. The compression test upon the diaphysis and the compact substance of the long bone of human extremities. *J. Kyoto Pref. Med. Univ.* 51:291–313.

Yoshikawa, K. 1964. Cleavage tests of human and animal compact bones. *J. Kyoto Pref. Med. Univ.* 73:121–134.

Yoshimatsu, N., H. Hazama, and T. Bando. 1956. Study on the extractive strength of the human teeth. *J. Kyoto Pref. Med. Univ.* 60:297–300.

Index

Printed and bound by CPI Group (UK) Ltd, Croydon, CR0 4YY

24/10/2024

01778307-0006